Protocol

LEONARDO

Roger F. Malina, series editor

Protocol

How Control Exists after Decentralization

Alexander R. Galloway

The MIT Press Cambridge, Massachusetts London, England

This book was set in Garamond and Bell Gothic by Graphic Composition, Inc., and
was printed and bound in the United States of America.

Library of Congress Cataloging-in-Publication Data

Galloway, Alexander R., 1974–
Protocol : how control exists after decentralization / Alexander R. Galloway.
p. cm.—(Leonardo)
Includes bibliographical references and index.
ISBN 978-0-262-07247-2 (hc. : alk. paper) — 978-0-262-57233-0 (pb. : alk. paper)
1. Computer networks—Security measures. 2. Computer networks—Management. 3. Computer network protocols. 4. Electronic data processing—Distributed processing. I. Title II. Leonardo (Series) (Cambridge, Mass.)

TK5105.59.G35   2004
005.8—dc22

2003060361

10 9 8 7 6 5 4

For Fredric Jameson and Michael Hardt, my teachers

Contents

Series Foreword

The cultural convergence of art, science, and technology provides ample opportunity for artists to challenge the very notion of how art is produced and to call into question its subject matter and its function in society. The mission of the Leonardo book series, published by The MIT Press, is to publish texts by artists, scientists, researchers, and scholars that present innovative discourse on the convergence of art, science, and technology.

Envisioned as a catalyst for enterprise, research, and creative and scholarly experimentation, the book series enables diverse intellectual communities to explore common grounds of expertise. The Leonardo book series provides a context for the discussion of contemporary practice, ideas, and frameworks in this rapidly evolving arena where art and science connect.

To find more information about Leonardo/ISAST and to order our publications, go to Leonardo Online at <http://mitpress.mit.edu/e-journals/Leonardo/isast/leobooks.html> or send e-mail to <leonardobooks@mitpress.mit.edu>.

Joel Slayton
Chair, Leonardo Book Series

Book Series Advisory Committee: Annick Bureaud, Pamela Grant Ryan, Craig Harris, Margaret Morse, Michael Punt, Douglas Sery, Allen Strange.
Leonardo/International Society for the Arts, Sciences, and Technology (ISAST)

Leonardo, the International Society for the Arts, Sciences, and Technology, and the affiliated French organization Association Leonardo have two very simple goals:

1. to document and make known the work of artists, researchers, and scholars interested in the ways that the contemporary arts interact with science and technology, and

2. to create a forum and meeting places where artists, scientists, and engineers can meet, exchange ideas, and, where appropriate, collaborate.

When the journal *Leonardo* was started some thirty-six years ago, these creative disciplines existed in segregated institutional and social networks, a situation dramatized at that time by the "Two Cultures" debates initiated by C. P. Snow. Today we live in a different time of cross-disciplinary ferment, collaboration and intellectual confrontation enabled by new hybrid organizations, new funding sponsors, and the shared tools of computers and the Internet. Above all, new generations of artist-researchers and researcher-artists are now at work individually and in collaborative teams bridging the art, science, and technology disciplines. Perhaps in our lifetime we will see the emergence of "new Leonardos," creative individuals or teams who will not only develop a meaningful art for our times but also drive new agendas in science and stimulate technological innovation that addresses today's human needs.

For more information on the activities of the Leonardo organizations and networks, please visit our Web site at <http://mitpress.mit.edu/Leonardo>.

Roger F. Malina
Chair, Leonardo/ISAST

Foreword: Protocol Is as Protocol Does

Eugene Thacker

There's a scene in the 1982 film *Tron,* in which Kevin Flynn, a computer programmer by day and hacker by night, gets sucked into the digital world of computers. There he meets software programs, each with personalities and identities of their own, held hostage by the "MCP" or "Master Control Program," a kind of despotic operating system that absorbs all software and all parts of the network into itself. One of the hostage programs named Tron, a security application leading a revolution against the MCP, asks Flynn about the mysteries of the world of "Users."

Flynn: I'm what you guys call a "User" . . .
Tron: Well if you are a user, then everything you've done has been according to a plan, right?
Flynn: (laughs) You wish . . . you guys know what it's like, you keep on doing what it looks like you're supposed to be doing, no matter how crazy it seems.
Tron: That's the way it is with programs, yes, but—
Flynn: I hate to disappoint you pal, but most of the time that's the way it is for users too . . .

Tron was made by Disney Studios in the early 1980s as part of an attempt to reinvent itself for a new generation of potential consumers. In particular, *Tron* was indicative of a particular type of early 1980s culture, one in which "personal" computers were becoming more and more ubiquitous, along with a booming video game industry and an equally large, middle-class geek culture to go with it. It was also a culture of perceived conflicts between computer mega-corporations and an emerging hacker subculture, both hatched in the midst of Reaganomics and the Japanese microelectronics boom. The list of gadgets spawned during this period has now entered the cultural museum of dead technology: video game arcades, Atari home consoles, Casio synthesizers, Commodore home computers, floppy disks, laserdiscs, the Walkman, and of course the neon lights of 1980s new wave music. It was from within this 1980s moment that William Gibson crafted the world of *Neuromancer,* and all of this is already making a comeback in the cultural nostalgia of "electroclash."

But more than being about fashion, the American middle-class youth culture *Tron* targeted was also one that existed during significant technological transformations, transformations we are still understanding today. The development of the personal computer, along with computer networks, has had a profound, stratified impact on the way in which social, political, and economic life is experienced. Recent discussions of the post-industrial society, the information society, the network society, disciplinary society, control society, informatization, scale-free networks, small worlds, and smart mobs are all ways of attempting to understand how social change is indissociable from technological development (research, design, use, distribution, marketing, naturalization, consumption)—though not determined by it. This last point is crucial. If one is to foster an understanding and awareness of how the social and the political are not external to technology, then it is important to understand how the technological is in some sense isomorphic to the social and the political.

This book—*Protocol*—points to one possible path for doing this. Throughout the discussions on power, control, and decentralization, *Protocol* consistently makes a case for a material understanding of technology. "Material" can be taken in all senses of the term, as an ontological category as well as a political and economic one. This type of materialist media studies shows how the question "how does it work?" is also the question "whom does it work for?" In short, the technical specs matter, ontologically and politically. As Galloway states here, "I attempt to read the never-ending stream of computer code as we read any text, decoding its structure of control." But this is clearly not literary criticism. Nor is it semiotics—textual, visual, or otherwise. This is because computer code is always enacted. Code is a set of procedures, actions, and practices, designed in particular ways to achieve particular ends in particular contexts. Code = praxis.

Protocol puts forth an invitation, a challenge to us: You have not sufficiently understood power relationships in the control society unless you have understood "how it works" and "who it works for." *Protocol* suggests that it is not only worthwhile, but also necessary to have a technical as well as theoretical understanding of any given technology. "Reading" code is thus more programming or development or debugging than explanation. In this sense, *Protocol* aims less to explain the society of control than to experiment with it; in fact, it might just as well be subtitled "experiments with code."

Which brings us to the following comments. *Protocol* implicitly makes three important points in its experiments with information networks, computer software, and industry standardization. The first has to do with how *Protocol* qualifies networks, the second point has to do with how *Protocol* understands the technical specs as political, and the last point looks toward possible future directions to be explored in the meeting of info-tech and bio-tech, info-politics and bio-politics.

Networks Are Real but Abstract

The first point is that networks are not metaphors. As Galloway states, "Protocols do not perform any interpretation themselves; that is, they encapsulate information inside various wrappers, while remaining relatively indifferent to the content of information contained within." The concept of "protocol" is thus meant to demonstrate the nonmetaphorical quality of networks. Or, put another way, the concept of protocol shows the predilection for general discussion of networks in terms of general tropes. Networks are not tropes for notions of "interconnection." They are material technologies, sites of variable practices, actions, and movements. This is, perhaps, stated too strongly. Yes, metaphors do materialize and corporealize, and, in some sense, metaphor is consonant with language itself. But discussions of networks—especially in cultural theory—have too often slipped into "vapor theory," eliding a specific consideration of the material substrate and infrastructure with a general discussion of links, webs, and globalized connectivity. "Protocol is a circuit, not a sentence." Further, code is not necessarily language, and certainly not a sign. A code, in the sense that *Protocol* defines it, is process-based: It is parsed, compiled, procedural or object-oriented, and defined by ontology standards. A code is a series of activated mechanical gears, or a stack of punched cards circulating through a tape-reading machine, or a flow of light-pulses or bits in a transistor or on silicon, or a vat of binding inter-actions between DNA fragments.

When the book suggests that networks are not metaphors (or not merely metaphors), the dichotomy is not one between material and immaterial, but rather between two types of "abstract." On the one hand there is an abstract use of the concept of networks generally to signify any relationship between discrete entities. According to this usage, just about anything can be seen as a network (and thus the overwhelmingly wide application of certain network science or complexity approaches). But there is also another meaning of "abstract," one that is not the opposite of concrete. An abstract that is real is a potential. (Henri Bergson uses the term "virtual" for the immanent unfolding of duration as potentiality.) This is not the abstract use of network as a term, and neither is it an abstraction of a technical term (the metaphorization of technological terms). Rather, this abstract-but-real is the network that is always enacted and always about to enact itself. One can pose the question: Is a network a network if it is not being used? Is the Internet a network because of its fiber-optic cables, its usage, its data transfer, the standards for such use, or the concepts that inform the development of network technology itself? Likely all of these. With multiple local agencies and several interests at stake, information networks like the Internet are always about to do something. In this sense networks are constantly materializing their logics (their formal logics as much as their organizational and political logics). The network as real-but-abstract may involve "information" as an immaterial entity, but that information always works toward real effects and transformations, no matter how localized.

Thus, in an important way, networks are not metaphors. The network metaphor is misleading, limiting. It only provides a general model for discussing relationships (linking, hybridity, boundaries, heterogeneity, etc.). As a cultural metaphor, networks only raise general issues of interrelationality. The discourse of cyberspace and the "information superhighway" is exemplary. Positions boil down to either the libertarian (information-wants-to-be-free) or the bureaucratic (gated access to information). Thus with the network metaphor, one can only see a nebulous thing called "information" that mysteriously exists on an equally nebulous thing called cyberspace or the Internet. Studying user interaction only adds to the confusion, bringing in the language of individual agency and accountability to a space that, to cite Jorge Luis Borges, has its center nowhere, its circumference everywhere.

Understanding networks not as metaphors, but as materialized and materializing media, is an important step toward diversifying and complexifying our understanding of power relationships in control societies. With the network metaphor, one only has a tool that does something in accordance to the agency of the human-user (a computer that downloads at your command, an information network that makes everything freely accessible at the click of a mouse, etc.). Click-download, cause-effect. If we dispense with convenient metaphors and actually ask how a network functions (not "what is it?" but "what does it do?"), then several noteworthy realizations emerge. This is what *Protocol* does. It asks how a particular type of network functions—the information networks that undergird the Internet. It shows how a network is not simply a free-for-all of information "out there," nor is it a dystopia of databanks owned by corporations. It is a set of technical procedures for defining, managing, modulating, and distributing information throughout a flexible yet robust delivery infrastructure. More than that, this infrastructure and set of procedures grows out of U.S. government and military interests in developing high-technology communications capabilities (from ARPA to DARPA to dot-coms). At an even finer level of detail, the Internet is not a simple "ask and you shall receive" tool. It is constituted by a bi-level logic that *Protocol* patiently explains. On the one hand, TCP/IP (Transmission Control Protocol/Internet Protocol) enables the Internet to create horizontal distributions of information from one computer to another. On the other, the DNS (Domain Name System) vertically stratifies that horizontal logic through a set of regulatory bodies that manage Internet addresses and names. Understanding these two dynamics in the Internet means understanding the essential ambivalence in the way that power functions in control societies. As *Protocol* states, "the founding principle of the Net is control, not freedom—control has existed from the beginning." To grasp "protocol" is to grasp the technical and the political dynamics of TCP/IP and DNS at the same time.

What are some common metaphoric uses of networks that *Protocol* resists? One is that networks are synonymous with connectivity (and that connectivity is always good). Tropes of links, nets, webs, and a general relationality also stem from this use of networks. Pop science writing on network science or the gee-whiz rhetoric of magazines such as *Wired* often adopt a quantitative politics: If information wants to be free, then more connectivity equals more freedom.

But the concept of "protocol" is as concerned with disconnection as it is with connectivity. The moment of disconnectivity is the moment when protocol most forcefully displays its political character. Disconnectivity can occur in a number of ways, from a personal dialup disconnected due to a time limit imposed by a commercial ISP, to the management (or surveillance) of large-scale networks involved in political activism. The analysis in part III below of hacking, computer viruses, and the tactical appropriation of both by cyberfeminism provides a set of examples for how disconnectivity can be as instrumental for protocol's (mal)functioning as is connectivity.

Aside from the trope of connectivity, another common metaphorical use of networks that *Protocol* resists is that of collectivity (and that collectivity is always inclusive). Here the language of incorporation, integration, and the constant struggle to include "the other" come together in discussions of virtual communities and online social services providing niche markets for every "difference."

But information protocols are always layered, stratified, sometimes blatantly hierarchical. *Protocol*'s analysis of the DNS system, and the management of "nodes" on the network, is exemplary in this case. The creation of standards is also the prescription of a range of practices that are legitimized (or delegitimized) within those standards. The example of the Name.Space project is instructive because it points to the ways in which both normativity as well as resistance are hardcoded into such standardization procedures. Again, the mere technical details, such as RFCs, suddenly become the grounds for contesting the way in which control takes shape in the materiality of networks.

To the common metaphors of connectivity and collectivity that are problematized, we can add a third, which is participation. Networks are participatory, even democratic (by nature). The notion of networks as participatory has led to much confusion regarding the status of individual and collective agency in such networks. This is the more prescient as the Web becomes a primary nexus of consumer culture, encompassing the production of commodities, information services, communication practices, and changing habits of consumption.

It is this naturalization of participation in networks that is particularly problematic. IRC, online gaming, blogs, e-registration, webcams, and online polls reiterate again and again the inherently participatory nature of the

Web. But one does not have to look far for instances in which some information was not meant to be free, instances in which participation is denaturalized, coded into secure servers, e-surveillance, predatory locales, and a new kind of gated community. The information surveillance practiced by the U.S. government on profiled Muslim individuals and groups is only the most explicit example. However, resistance is built in. The examples of "software art," open source activities, and network politics (such as the Electronic Disturbance Theater) all provide examples of potentials for "counterprotocol" practices.

A recent case helps to show how the vague utilization of network metaphors (connectivity, collectivity, participation) can be problematized in specific contexts. An example is The Thing, a New York-based, nonprofit ISP and virtual community dedicated to art, politics, and the open discussion of a range of cultural and political issues. On December 4, 2002, The Thing's connection was interrupted by Verio, its network provider, on the grounds that The Thing had infringed its service contract with Verio. The contract was terminated a short while later, leaving The Thing, its members, participants, and some two hundred projects in limbo. According to Verio, the perceived cause of the disconnection was an art-activist group known as The Yes Men. On the eighteenth anniversary of the Bhopal disaster, in which a Union Carbide/Dow chemical accident resulted in the illness and death of thousands of Bhopal citizens, The Yes Men had circulated a fabricated press release from the Dow corporation disavowing any responsibility for the accident. The fake press release was intended not only to continue to raise awareness of such incidents, but to raise issues concerning ongoing discourses of globalism and corporate management of information networks. Dow, notified of the fake press release, filed a DMCA (Digital Millennium Copyright Act) notice to Verio. Verio temporarily, and then permanently, pulled the plug on The Thing, as the ISP hosting The Yes Men's website.

Is this an instance of censorship? Is it protocol malfunctioning or functioning too well? Do the politics of this disconnectivity affect the Bhopal community itself? Do policies regarding information content (DMCA) actually encourage such regulation? These are complex questions which have arisen from this event, one that has been made possible by the dual nature of the Internet that *Protocol* points to: its horizontality (community networks; TCP/IP) and its verticality (its stratification; DNS).

Protocol, or Political Economy

If, in the discourses surrounding networks, the tropes of connectivity, collectivity, and participation obscure the material practices of networks, *Protocol* points to several principles for understanding networks as "a diagram, a technology, and a management style." To begin with, general talk about "networks," dissociated from their context and technical instantiation, can be replaced by a discussion of "protocols." Every network is a network because it is constituted by a protocol. If networks display any of the tropes described previously, it is because there is an infrastructure that enables such properties to emerge. Not networks, but protocols.

Given this, *Protocol* can be read as a book of political economy. It argues for a methodological shift from a generalized understanding of networks to a specified one, in which the protocological systems of TCP/IP and DNS operate as what Foucault termed "political technologies." Foucault's later work on biopower and biopolitics is significant in this respect, for while Foucault never reduced technology to an empirical "thing," his analyses of institutions always emphasize the various correlations between bodies and things. *Protocol* adopts a similar methodological outlook, considering technical standards (such as the OSI Reference Model), network technologies (HTTP), institutional histories (IEEE), and, significantly, instances of "tactical media."

Above all, the political economy of protocol is that of management, modulation, and control. Technically and politically the "control society" emerges as much from cybernetics research as it does from a military-industrial imperative toward the "governmentality" of information systems. This historical backdrop sets the stage for the various periodizations and mutations in the life of protocol. At the center of such changes is the issue of political and technological adaptation, situated between centralized control and decentralized regulation.

As a political economy, protocol modulates, regulates, and manages the interrelationality between vital systems. In this sense, a "vital system" is not just a living, biological system, nor is it nineteenth-century "vitalism," or worse, animism. Vital systems have to be understood from the perspective of protocological control. While it may be a truism that the body stands in relation to, but never identical with, technology, protocological control makes different cuts. *Protocol* considers networks through a "diagram," a term borrowed from Gilles Deleuze. *Protocol* considers first a network as a set of nodes and edges, dots and lines. The dots may be computers (server, client, or both), human users, communities, LANs, corporations, even countries. The lines can be any practice, action, or event effectuated by the dots (downloading, emailing, connecting, encrypting, buying, logging on, port scanning). With this basic "diagram" you can do a number of things. You can connect the dots—all of them—making a totally connected, distributed network with more than one path to the destination. You can also disconnect dots, even delete dots (no paths, no destination). You can filter out which dots are connected to the network. You can create portals for the addition of future dots. You can designate which kinds of lines you want between the dots (for not all lines are equal; some diverge, flee; others converge, coalesce). In short, a network-as-diagram offers all sorts of possibilities for organization, regulation, and management.

But this is depending, of course, on the agencies responsible for the network-as-diagram. As *Protocol* makes clear, there are few instances in which a clearly demarcated, centralized network control is evident. Paraphrasing Foucault, such instances occur only at the terminal ends of power relationships. The central political question that *Protocol* asks is where the power has gone. If we are indeed living in a post-industrial, postmodern, postdemocratic society, how does one account for political agency in situations in which agency appears to be either caught in networks of power or distributed across multiple agencies?

By looking closely and carefully at the technical specifications of TCP/IP and DNS, *Protocol* suggests that power relations are in the process of being transformed in a way that is resonant with the flexibility and constraints of information technology. The Internet is not simply "open" or "closed" but above all a form that is modulated. The very concept of packet-switching demonstrates this on several levels, from the efficiency standards of routing during a download, to the ways in which each individual datagram is tagged for delivery to your email account or hard drive. Information does flow, but it does so in a highly regulated manner. This dual property (regulated flow) is central to *Protocol*'s analysis of the Internet as a political technology.

Isomorphic Biopolitics

As a final comment, it is worthwhile to note that the concept of "protocol" is related to a biopolitical production, a production of the possibility for experience in control societies. It is in this sense that *Protocol* is doubly materialist—in the sense of networked bodies inscribed by informatics, and in the sense of this bio-informatic network producing the conditions of experience.

The biopolitical dimension of protocol is one of the parts of this book that opens onto future challenges. As the biological and life sciences become more and more integrated with computer and networking technology, the familiar line between the body and technology, between biologies and machines, begins to undergo a set of transformations. "Populations" defined nationally or ethnically are also defined informatically. (Witness the growing business of population genomics.) Individual subjects are not only civil subjects, but also medical subjects for a medicine increasingly influenced by genetic science. The ongoing research and clinical trials in gene therapy, regenerative medicine, and genetic diagnostics reiterate the notion of the biomedical subject as being in some way amenable to a database. In addition to this bio-informatic encapsulation of individual and collective bodies, the transactions and economies between bodies are also being affected. Research into stem cells has ushered in a new era of molecular bodies that not only are self-generating like a reservoir (a new type of tissue banking), but that also create a tissue economy of potential biologies (lab-grown tissues and organs).

Such biotechnologies often seem more science fiction than science, and indeed health care systems are far from fully integrating such emerging research into routine medical practice. In addition, this seems to be far from the "dry" world of bits and data transfer. So then, what is the relation between protocol and biopolitics?

One response is that protocol is isomorphic with biopolitics. Another way of saying the same thing is that "information" is often taken to be isomorphic with vital forms. There is an uneasy dialectic here between a living computer (artificial life) and the programming of life (genetic engineering). From the perspective of protocol, the nature/culture, body/technology binarisms do not matter. Literally. Rather, what matters is the ability of protocol to operate across material-energetic substrates. This, in itself, is not "bad," and as *Protocol* suggests, the question is not one of morality, but rather of ethics. Interests are at stake. From the perspective of protocol, there are no biologies, no technologies, only the possible interactions between "vital forms" which often take on a regulatory, managerial, and normative shape. This can be called biopolitics. In the context of protocol, the reach between info-tech and bio-tech is enabled by several technical concepts.

Layering is a central concept of the regulation of information transfer in the Internet protocols. Layering allows data to find its way into the correct application on your computer, so that an MP3 download is not misunderstood as a Word document, or a virus is not mistaken as a browser plug-in. A datagram coming into your computer moves from the lower levels (encoding of packets) to the higher levels (use by applications). Further, *Protocol* suggests that there exists a layering between the biological and the political. A signaling pathway between two proteins is layered into a disease predisposition pathway in a genome, which is layered onto a genome database, which is layered onto a statistical and demographic record of disease occurrence, which is layered onto research funds going toward biowarfare detection technologies, which is layered into popular science journalism or even science fiction film. Note that some of these layerings are more metaphorical, while others are almost exclusively technical. Sometimes the layering is quite sedimented, there being only one way in, one way out. At other times the layering shifts, rearranges itself, nature not necessarily preceding culture, culture not necessarily preceding nature.

Portability is a central characteristic of software development. Mac or PC? Netscape or IE? The ability to enable software and files to operate across different proprietary standards is a key aspect of software development. In a sense, layering cannot happen without at least a cursory consideration of portability. Portability is not always the will to make something portable; more often than not it is the strategic disabling of portability that is of concern to software companies. If the biological body is a genetic code, and if the genome is a kind of computer, then it follows that the main area of interest for portability will be between the biological body of the patient and the informatic body of the computer database or profile. Despite the ongoing discussion over cloning or gene therapy, some suggest that it will be diagnostic tools that guarantee financial sustainability for the biotech and pharmaceutical industries. The key to that success will be the portability between the two types of genetic codes: one in vivo, one in silico.

Ontology standards is a strange name for agreed-upon code conventions, but in some circles it is regularly used to signify just that. Newer, more flexible markup languages such as XML (Extensible Markup Language) have made it possible for researchers (be they biologists or engineers) to come up with a coding schema tailored to their discipline. XML-based efforts in molecular biology and biochemistry have been one area of concern. But agreeing upon what exactly that standard code will be is another matter. Should the hierarchy of tags for GEML (Gene Expression Markup Language) go by <chromosome>, <phenotype>, or <gene>? There are a range of vested interests (commercial, ideological, institutional, methodological, discipli-nary), and the mere decision about standards becomes a discourse on "ontol-ogy" in the philosophical sense. If layering is dependent upon portability, then portability is in turn enabled by the existence of ontology standards.

These are some of the sites that *Protocol* opens up concerning the possible relations between information and biological networks. While the concept of biopolitics is often used at its most general level, *Protocol* asks us to re-specify biopolitics in the age of biotechnology and bioinformatics. Thus one site of future engagement is in the zones where info-tech and bio-tech inter-sect. The "wet" biological body has not simply been superceded by "dry" computer code, just as the wet body no longer accounts for the virtual body. Biotechnologies of all sorts demonstrate this to us—in vivo tissue engineer-ing, ethnic genome projects, gene-finding software, unregulated genetically modified foods, portable DNA diagnostics kits, and distributed proteomic computing. Protocological control in biopolitical contexts is not just the means, but the medium for the development of new forms of management, regulation, and control.

On a general level, *Protocol* provides a set of concepts, or a toolbox, to use Deleuze's phrase. These concept-tools are not so much instruments or ham-mers, but rather soft machines for interrogating the political-technical dy-namics of information and computer technologies, especially as they pertain to networks of all types. *Protocol* can in this sense be read as a technical man-ual, one that fosters working with, intervening in, and building awareness of our current "political technologies." This is the kind of book that asks us not to interpret, but to experiment.

Preface

You may be wondering why someone with a cultural and literary background, not a scientific one, is writing a book on computer protocols. Over the last few years, I have worried over and marveled at this very possibility. Discipline-hopping is a necessity when it comes to complicated sociotechnical topics like protocol. As Michael Hardt and Kathi Weeks write in their introduction to *The Jameson Reader,* "[w]e have to delve into other disciplines and other realms of social production, not merely as a hobby or a supplement to our primary work, but as an integral part of it."[1] I take their advice very seriously.

So this is a book on computer science written by someone whose training is not in computer science. While I have some experience as a systems administrator and computer programmer, I am, admittedly, an amateur. My formal training is in critical theory—from cultural studies, film theory, psychoanalysis, and feminism to Marxism, semiotics, structuralism, and post-structuralism. Perhaps it was Gilles Deleuze who first convinced me that one can remain a theorist and still spend much of one's time talking about science. This remains the conundrum of interdisciplinary work: to do the work of another scholarly field, but not to be that field's scholar. And so I offer this book as an example of fruitful research brought to one field by a specialist in another.

Like film was to André Bazin or fashion was to Roland Barthes, I consider computers to be fundamentally a textual medium. The reason is obvious:

---

1. Michael Hardt and Kathi Weeks, eds., *The Jameson Reader* (Oxford: Blackwell, 2000), p. 2.

computers are based on a technological language called code. This underlying code provides the textual link between computers and critical theory.

Media critic Friedrich Kittler has noted that in order for one to understand contemporary culture, one must understand at least one natural language and at least one computer language. It is my position that the largest oversight in contemporary literary studies is the inability to place computer languages on par with natural languages—something I address in chapter 5 below on hacking.

There are, I admit, important historical and sociological reasons for not doing so. For example, natural languages like English or Latin are much older than computer languages. But I would suggest that there are few logical explanations for excluding computer discourse from the fields of literary criticism, linguistics, or even poetry.

Computer languages exhibit many of the qualities that define natural languages. Like the natural languages, they have their own sophisticated syntax and grammar. Like the natural languages, they exist in specific communities and cultures, uniting a community through shared meaning and value.

For all these reasons, I find my evolution out of literature, critical theory, and film into new media to be a very natural step indeed. I hope literature specialists, philosophers, scientists, and technophobes alike will think so too.

Acknowledgments

This work follows directly from my doctoral thesis. I would like to thank Fredric Jameson, Michael Hardt, Jane Gaines, and Lawrence Grossberg who sat on my dissertation committee and provided invaluable support along the way. Other teachers who influenced me include Neil Lazarus, Ellen Rooney, Philip Rosen, Michael Silverman, and Elizabeth Weed. Anne Stephens and Jim Wichterman sparked my curiosity at a young age.

I have been influenced and assisted by many people including Steve Cisler, Ricardo Dominguez, Anne Ellegood, Ken Goldberg, Branden Hookway, Jon Ippolito, Natalie Jeremijenko, Thomas Levin, Geert Lovink, Lev Manovich, Milton Mueller, Jonah Peretti, and McKenzie Wark. Friend and collaborator Eugene Thacker graciously agreed to write the foreword. Rachel Greene coauthored several texts with me in 1998, scraps of which reappear, however transformed, in various parts of chapter 7. The following people agreed to be interviewed by me: Bernard Aboba, Fred Baker, Bob Braden, Vinton Cerf, Steve Crocker, Stacy Leistner (ANSI), Tony Rutkowski, and Stefaan Verhulst. Kristen DelGrosso provided invaluable research assistance for portions of this book. I would also like to thank Doug Sery at the MIT Press and series editor Joel Slayton for their support and assistance.

I thank my family for their love and support, especially my brother Munro Galloway, who has, over the years, generally been first to introduce me to the more interesting areas of culture.

For my technical training I thank Rhizome and my partner there, Mark Tribe. My six years at Rhizome allowed me to learn the pleasures of relational databases, Web-based software applications, general Linux system administration—and of course new media art—in an exciting, creative environment.

Thanks to Mark Amerika who organized the exhibition Digital Studies with me in 1997. The short text I wrote for that show—"What Is Digital Studies?"—became the germ of inspiration for this entire book. Likewise, Arthur and Marilouise Kroker at *CTHEORY* helped kick-start this book by publishing a very early piece of mine.

Last but not least I am indebted to the new media communities and organizations that exist in various areas of virtual and real space: *Rhizome, Nettime, 7-11/American Express, CTHEORY,* Ars Electronica, Next 5 Minutes (Society for Old and New Media), Eyebeam, ZKM, V2, and many, many others. Without them I would never have been able to figure out what to say in this book.

I

How Control Exists After Decentralization

Introduction

Every society has its diagram(s).
—GILLES DELEUZE, *Foucault*

This book is about a diagram, a technology, and a management style. The diagram is the *distributed network,* a structural form without center that resembles a web or meshwork. The technology is the digital *computer,* an abstract machine able to perform the work of any other machine (provided it can be described logically). The management style is *protocol,* the principle of organization native to computers in distributed networks. All three come together to define a new apparatus of control that has achieved importance at the start of the new millennium.

Much work has been done recently on theorizing the present historical moment and on offering periodizations to explain its historical trajectory. I am particularly inspired by five pages from Gilles Deleuze, "Postscript on Control Societies," which begin to define a chronological period after the modern age that is founded neither on the central control of the sovereign nor on the decentralized control of the prison or the factory. My book aims to flesh out the specificity of this third historical wave by focusing on the controlling computer technologies native to it.

How would control exist after decentralization? In former times control was a little easier to explain. In what Michel Foucault called the sovereign societies of the classical era, characterized by centralized power and sovereign fiat, control existed as an extension of the word and deed of the master, assisted by violence and other coercive factors. Later, the disciplinary societies of the modern era took hold, replacing violence with more bureaucratic forms of command and control.

Deleuze has extended this periodization into the present day by suggesting that after the disciplinary societies come the *societies of control.* Deleuze believed that there exist wholly new technologies concurrent with the societies of control. "The old sovereign societies worked with simple machines, levers, pulleys, clocks," he writes, "but recent disciplinary societies were equipped with thermodynamic machines[1] . . . control societies operate with a third generation of machines, with information technology and

---

Epigraph: Gilles Deleuze, *Foucault,* trans. Seán Hand (Minneapolis: University of Minnesota Press, 1986), p. 35.

1. "Thermodynamic machines" refers primarily to steam and internal combustion engines and to nuclear power.

computers."[2] Just as Marx rooted his economic theory in a strict analysis of the factory's productive machinery, Deleuze heralds the coming productive power of computers to explain the sociopolitical logics of our own age.

According to Critical Art Ensemble (CAE), the shift from disciplinary societies to control societies goes something like this:

> Before computerized information management, the heart of institutional command and control was easy to locate. In fact, the conspicuous appearance of the halls of power was used by regimes to maintain their hegemony. . . . Even though the monuments of power still stand, visibly present in stable locations, the agency that maintains power is neither visible nor stable. Power no longer permanently resides in these monuments, and command and control now move about as desired.[3]

The most extensive "computerized information management" system existing today is the Internet. The Internet is a global distributed computer network. It has its roots in the American academic and military culture of the 1950s and 1960s.[4] In the late 1950s, in response to the Soviet Sputnik launch and other fears connected to the Cold War,[5] Paul Baran at the Rand Corpo-

---

2. Gilles Deleuze, "Postscript on Control Societies," in *Negotiations,* trans. Martin Joughin (New York: Columbia University Press, 1990), p. 180; an alternate translation is available as "Postscript on the Societies of Control" in *October: The Second Decade, 1986–1996,* ed. Rosalind Krauss et al. (Cambridge: MIT Press, 1997).

3. Critical Art Ensemble, *Electronic Civil Disobedience and Other Unpopular Ideas* (New York: Autonomedia, 1996), pp. 7–8, 9.

4. Katie Hafner and Matthew Lyon dispute this in their book *Where Wizards Stay Up Late: The Origins of the Internet* (New York: Touchstone, 1996), arguing instead that the Internet was derived from the altruistic concerns of a few academics rather than the strategic interests of the Department of Defense. Yet they equivocate, writing on the one hand that "[t]he project had embodied the most peaceful intentions—to link computers at scientific laboratories across the country so that researchers might share computer resources. . . . the ARPANET and its progeny, the Internet, had nothing to do with supporting or surviving war—never did" (p. 10); yet on the other hand they admit that Paul Baran, the man who has contributed most to the emergence of protocol, "developed an interest in the survivability of communications systems under nuclear attack" (p. 54).

5. American anxiety over Soviet technological advancement was very real after the Sputnik launches. "The launching of the sputniks told us," wrote John Dunning for *The New York Times*

ration decided to create a computer network that was independent of centralized command and control, and would thus be able to withstand a nuclear attack that targets such centralized hubs. In August 1964, he published an eleven-volume memorandum for the Rand Corporation outlining his research.[6]

Baran's network was based on a technology called packet-switching[7] that allows messages to break themselves apart into small fragments. Each fragment, or packet, is able to find its own way to its destination. Once there, the packets reassemble to create the original message. In 1969, the Advanced Research Projects Agency (ARPA) at the U.S. Department of Defense started the ARPAnet, the first network to use Baran's packet-switching technology. The ARPAnet allowed academics to share resources and transfer files. In its early years, the ARPAnet (later renamed DARPAnet) existed unnoticed by the outside world, with only a few hundred participating computers, or "hosts."

All addressing for this network was maintained by a single machine located at the Stanford Research Institute in Menlo Park, California. By 1984 the network had grown larger. Paul Mockapetris invented a new addressing scheme, this one decentralized, called the Domain Name System (DNS).

The computers had changed also. By the late 1970s and early 1980s personal computers were coming to market and appearing in homes and offices. In 1977, researchers at Berkeley released the highly influential "BSD" flavor of the UNIX operating system, which was available to other institutions at

---

Magazine in 1957, "that a great despotism is now armed with rockets of enormous thrust, and guidance systems that could deliver a hydrogen warhead of one or more megatons to any spot in the United States." See John Dunning, "If We Are to Catch Up in Science," *New York Times Magazine*, November 10, 1957, p. 19.

6. Baran tells us that these memoranda "were primarily written on airplanes in the 1960 to 1962 era." See Paul Baran, Electrical Engineer, an oral history conducted in 1999 by David Hochfelder, IEEE History Center, Rutgers University, New Brunswick, NJ, USA.

7. A term coined instead by British scientist Donald Davies who, unknowing of Baran's work, also invented a system for sending small packets of information over a distributed network. Both scientists are credited with the discovery; however, because of Baran's proximity to the newly emerging ARPA network, which would be the first to use Baran's ideas, Davies's historical influence has diminished.

virtually no cost. With the help of BSD, UNIX would become the most important computer operating system of the 1980s.

In the early 1980s, the suite of protocols known as TCP/IP (Transmission Control Protocol/Internet Protocol) was also developed and included with most UNIX servers. TCP/IP allowed for cheap, ubiquitous connectivity. In 1988, the Defense department transferred control of the central "backbone" of the Internet over to the National Science Foundation, who in turn transferred control to commercial telecommunications interests in 1995. In that year, there were 24 million Internet users. Today, the Internet is a global distributed network connecting billions of people around the world.

At the core of networked computing is the concept of *protocol*. A computer protocol is a set of recommendations and rules that outline specific technical standards. The protocols that govern much of the Internet are contained in what are called RFC (Request For Comments) documents.[8] Called "the primary documentation of the Internet,"[9] these technical memoranda detail the vast majority of standards and protocols in use on the Internet today.

The RFCs are published by the Internet Engineering Task Force (IETF). They are freely available and used predominantly by engineers who wish to build hardware or software that meets common specifications. The IETF is affiliated with the Internet Society, an altruistic, technocratic organization that wishes "[t]o assure the open development, evolution and use of the Internet for the benefit of all people throughout the world."[10] Other protocols are developed and maintained by other organizations. For example, many of the protocols used on the World Wide Web (a network within the Internet) are governed by the World Wide Web Consortium (W3C). This international consortium was created in October 1994 to develop common protocols such as Hypertext Markup Language (HTML) and Cascading Style Sheets. Scores of other protocols have been created for a variety of other purposes by many

---

8. The expression derives from a memorandum titled "Host Software" sent by Steve Crocker on April 7, 1969, which is known today as RFC 1.

9. Pete Loshin, *Big Book of FYI RFCs* (San Francisco: Morgan Kaufmann, 2000), p. xiv.

10. "Internet Society Mission Statement," available online at http://www.isoc.org/isoc/mission/.

different professional societies and organizations. They are covered in more detail in chapter 4.

Protocol is not a new word. Prior to its usage in computing, protocol referred to any type of correct or proper behavior within a specific system of conventions. It is an important concept in the area of social etiquette as well as in the fields of diplomacy and international relations. Etymologically it refers to a fly-leaf glued to the beginning of a document, but in familiar usage the word came to mean any introductory paper summarizing the key points of a diplomatic agreement or treaty.

However, with the advent of digital computing, the term has taken on a slightly different meaning. Now, protocols refer specifically to standards governing the implementation of specific technologies. Like their diplomatic predecessors, computer protocols establish the essential points necessary to enact an agreed-upon standard of action. Like their diplomatic predecessors, computer protocols are vetted out between negotiating parties and then materialized in the real world by large populations of participants (in one case citizens, and in the other computer users). Yet instead of governing social or political practices as did their diplomatic predecessors, computer protocols govern how specific *technologies* are agreed to, adopted, implemented, and ultimately used by people around the world. What was once a question of consideration and sense is now a question of logic and physics.

To help understand the concept of computer protocols, consider the analogy of the highway system. Many different combinations of roads are available to a person driving from point A to point B. However, en route one is compelled to stop at red lights, stay between the white lines, follow a reasonably direct path, and so on. These conventional rules that govern the set of possible behavior patterns within a heterogeneous system are what computer scientists call protocol. Thus, protocol is a technique for achieving voluntary regulation within a contingent environment.

These regulations always operate at the level of coding—they encode packets of information so they may be transported; they code documents so they may be effectively parsed; they code communication so local devices may effectively communicate with foreign devices. Protocols are highly formal; that is, they encapsulate information inside a technically defined wrapper, while remaining relatively indifferent to the content of information contained within. Viewed as a whole, protocol is a distributed management system that allows control to exist within a heterogeneous material milieu.

It is common for contemporary critics to describe the Internet as an unpredictable mass of data—rhizomatic and lacking central organization. This position states that since new communication technologies are based on the elimination of centralized command and hierarchical control, it follows that the world is witnessing a general disappearance of control as such.

This could not be further from the truth. I argue in this book that protocol is how technological control exists after decentralization. The "after" in my title refers to both the historical moment after decentralization has come into existence, but also—and more important—the historical phase *after* decentralization, that is, after it is dead and gone, replaced as the supreme social management style by the diagram of distribution.

What contributes to this misconception (that the Internet is chaotic rather than highly controlled), I suggest, is that protocol is based on a *contradiction* between two opposing machines: One machine radically distributes control into autonomous locales, the other machine focuses control into rigidly defined hierarchies. The tension between these two machines—a dialectical tension—creates a hospitable climate for protocological control.

Emblematic of the first machinic technology, the one that gives the Internet its common image as an uncontrollable network, is the family of protocols known as TCP/IP. TCP and IP are the leading protocols for the actual transmission of data from one computer to another over the network. TCP and IP work together to establish connections between computers and move data packets effectively through those connections. Because of the way TCP/IP was designed, any computer on the network can talk to any other computer, resulting in a nonhierarchical, peer-to-peer relationship.

As one technical manual puts it: "IP uses an anarchic and highly distributed model, with every device being an equal peer to every other device on the global Internet."[11] (That a technical manual glowingly uses the term "anarchic" is but one symptom of today's strange new world!)

Emblematic of the second machinic technology, the one that focuses control into rigidly defined hierarchies, is the DNS. DNS is a large decentralized

---

11. Eric Hall, *Internet Core Protocols: The Definitive Guide* (Sebastopol, CA: O'Reilly, 2000), p. 407.

database that maps network addresses to network names. This mapping is required for nearly every network transaction. For example, in order to visit "www.rhizome.org" on the Internet one's computer must first translate the name "www.rhizome.org," itself geographically vague, into a specific address on the physical network. These specific addresses are called IP addresses and are written as a series of four numbers like so: 206.252.131.211.

All DNS information is controlled in a hierarchical, inverted-tree structure. Ironically, then, nearly all Web traffic must submit to a hierarchical structure (DNS) to gain access to the anarchic and radically horizontal structure of the Internet. As I demonstrate later, this contradictory logic is rampant throughout the apparatus of protocol.

The process of converting domain names to IP addresses is called *resolution*. At the top of this inverted tree are a handful of so-called "root" servers holding ultimate control and delegating lesser control to lower branches in the hierarchy. There are over a dozen root servers located around the world in places like Japan and Europe, as well as in several U.S. locations.

To follow the branches of control, one must parse the address in reverse, starting with the top-level domain, in this case "org." First, the root server receives a request from the user and directs the user to another machine that has authority over the "org" domain, which in turn directs the user to another machine that has authority over the "rhizome" subsection, which in turn returns the IP address for the specific machine known as "www."

Only the computer at the end of the branch knows about its immediate neighborhood, and thus it is the only machine with authoritative DNS information. In other words resolution happens like this: A new branch of the tree is followed at each successive segment, allowing the user to find the authoritative DNS source machine and thus to derive the IP address from the domain name. Once the IP address is known, the network transaction can proceed normally.

Because the DNS system is structured like an inverted tree, each branch of the tree holds absolute control over everything below it. For example, in the winter of 1999, a lawsuit was brought against the Swiss art group Etoy. Even though the basis of the lawsuit was questionable and was later dropped, the courts would have been able to "turn off" the artist's Web site during the course of the trail by simply removing DNS support for "etoy.com." (Instead the artists were forced to pull the plug themselves until after the trial was over.)

A similar incident happened at The Thing, an Internet service provider based in New York who was hosting some of Etoy's agitprop. After some of this material was deemed politically questionable by the Federal Bureau of Investigation, the whole server was yanked off the Internet by the telecommunications company who happened to be immediately upstream from the provider. The Thing had no recourse but to comply with this hierarchical system of control.

The inventor of the World Wide Web, Tim Berners-Lee, describes the DNS system as the "one centralized Achilles' heel by which [the Web] can all be brought down or controlled."[12]

If hypothetically some controlling authority wished to ban China from the Internet (e.g., during an outbreak of hostilities), they could do so very easily through a simple modification of the information contained in the root servers at the top of the inverted tree. Within twenty-four hours, China would vanish from the Internet.

As DNS renegade and Name.Space founder Paul Garrin writes: "With the stroke of a delete key, whole countries can be blacked out from the rest of the net. With the "." [root file] centralized, this is easily done. . . . Control the "." and you control access."[13] Since the root servers are at the top, they have ultimate control over the existence (but not necessarily the content) of each lesser branch. Without the foundational support of the root servers, all lesser branches of the DNS network become unusable. Such a reality should shatter our image of the Internet as a vast, uncontrollable meshwork.

Any networked relation will have multiple, nested protocols. To steal an insight from Marshall McLuhan, *the content of every new protocol is always another protocol.* Take, for example, a typical transaction on the World Wide Web. A Web page containing text and graphics (themselves protocological artifacts) is marked up in the HTML protocol. The protocol known as Hypertext Transfer Protocol (HTTP) encapsulates this HTML object and allows it to be served by an Internet host. However, both client and host must abide by the TCP protocol to ensure that the HTTP object arrives in one piece. Finally, TCP is itself nested within the Internet Protocol, a protocol

---

12. Tim Berners-Lee, *Weaving the Web* (New York: HarperCollins, 1999), p. 126.

13. Paul Garrin, "DNS: Long Winded and Short Sighted," *Nettime,* October 19, 1998.

that is in charge of actually moving data packets from one machine to another. Ultimately the entire bundle (the primary data object encapsulated within each successive protocol) is transported according to the rules of the only "privileged" protocol, that of the physical media itself (fiber-optic cables, telephone lines, air waves, etc.). The flexible networks and flows identified in the world economy by Manuel Castells and other anchormen of the Third Machine Age are not mere metaphors; they are in fact built directly into the technical specifications of network protocols. By design, protocols such as the Internet Protocol *cannot be centralized.*

Protocol's native landscape is the distributed network. Following Deleuze, I consider the distributed network to be an important *diagram* for our current social formation. Deleuze defines the diagram as "a map, a cartography that is coextensive with the whole social field."[14] The distributed network is such a map, for it extends deeply into the social field of the new millennium. (I explore this point in greater detail in chapter 1.)

A distributed network differs from other networks such as centralized and decentralized networks in the arrangement of its internal structure. A centralized network consists of a single central power point (a host), from which are attached radial nodes. The central point is connected to all of the satellite nodes, which are themselves connected only to the central host. A decentralized network, on the other hand, has *multiple* central hosts, each with its own set of satellite nodes. A satellite node may have connectivity with one or more hosts, but not with other nodes. Communication generally travels unidirectionally within both centralized and decentralized networks: from the central trunks to the radial leaves.

The distributed network is an entirely different matter. Distributed networks are native to Deleuze's control societies. Each point in a distributed network is neither a central hub nor a satellite node—there are neither trunks nor leaves. The network contains nothing but "intelligent end-point systems that are self-deterministic, allowing each end-point system to communicate with any host it chooses."[15] Like the rhizome, each node in a distributed network may establish direct communication with another node,

---

14. Deleuze, *Foucault,* p. 34.

15. Hall, *Internet Core Protocols,* p. 6.

without having to appeal to a hierarchical intermediary. Yet in order to initiate communication, the two nodes must *speak the same language.* This is why protocol is important. Shared protocols are what defines the landscape of the network—who is connected to whom.

As architect Branden Hookway writes: "[d]istributed systems require for their operation a homogenous standard of interconnectivity."[16] Compatible protocols lead to network articulation, while incompatible protocols lead to network disarticulation. For example, two computers running the DNS addressing protocol will be able to communicate effectively with each other about network addresses. Sharing the DNS protocol allows them to be networked. However, the same computers will not be able to communicate with foreign devices running, for example, the NIS addressing protocol or the WINS protocol.[17] Without a shared protocol, there is no network.

I turn now to Michel Foucault to derive one final quality of protocol, the special existence of protocol in the "privileged" physical media of *bodies.* Protocol is not merely confined to the digital world. As Deleuze shows in the "Postscript on Control Societies," protocological control also affects the functioning of bodies within social space and the creation of these bodies into forms of "artificial life" that are *dividuated,*[18] sampled, and coded. "Artificial life" is a term I use in chapter 3 to describe protocol *within the sociopolitical theater.* Artificial life simply means the active production of vital forms by other vital forms—what Foucault calls the "work of the self on the self."

I later suggest that Foucault's relationship to life forms is a protocological one. This is expressed most clearly in his later work, particularly in the twin concepts of biopolitics and biopower. Foucault defines biopolitics as "the endeavor, begun in the eighteenth century, to rationalize the problems presented to governmental practice by the phenomena characteristic of a

---

16. Branden Hookway, *Pandemonium: The Rise of Predatory Locales in the Postwar World* (New York: Princeton Architectural Press, 1999), p. 77.

17. WINS, or Windows Internet Name Service, is an addressing technology developed by Microsoft for distributed networks; NIS, or Network Information Service, is a similar technology developed by Sun Microsystems.

18. Deleuze's neologism comes from the word "*in*dividuate." Dividuation would thus be the opposite: the dissolving of individual identity into distributed networks of information.

group of living human beings constituted as a population: health, sanitation, birthrate, longevity, race."[19] Thus one can assume that technologies like biometrics and statistical analysis—from the Bertillon identification system, to the Social Security Act of 1935, to the tabulation of birth rates by the Children's Defense Fund—all fall into the category biopolitics.

Further, he writes that biopolitics "tends to treat the 'population' as a mass of living and coexisting beings who present particular biological and pathological traits and who thus come under specific knowledge and technologies."[20] Biopolitics, then, connects to a certain statistical knowledge about populations. It is a species-knowledge (an expression that sounds less ominous if one considers an allusion to Marx's utopian concept of "species-being").

Still, Foucault puts equal stress on "technologies" and "knowledge" in his definition of biopolitics. But which technologies in particular would correspond to Foucault's biopolitical scenario? I argue here that they are the distributed forms of management that characterize the contemporary computer network and within which protocological control exists.

In *The History of Sexuality, Volume 1,* Foucault contrasts the older power of the sovereign over life (one characterized by the metaphysical concern of either the absence or presence of life) to a new mode in which life is either created or destroyed: "One might say that the ancient right to *take* life or *let* live was replaced by a power to *foster* life or *disallow* it to the point of death."[21] He continues: "The old power of death that symbolized sovereign power was now carefully supplanted by the *administration of bodies* and the *calculated management of life.*"[22] Foucault's treatment of biopower is entirely protocological. Protocol is to control societies as the panopticon is to disciplinary societies.

While protocol may be more *democratic* than the panopticon in that it strives to eliminate hierarchy, it is still very much structured around command and control and therefore has spawned counter-protocological forces.

---

19. Michel Foucault, *Ethics: Subjectivity and Truth,* ed. Paul Rabinow (New York: New Press, 1997), p. 73.

20. Foucault, *Ethics,* p. 71.

21. Michel Foucault, *The History of Sexuality, Volume 1,* trans. Robert Hurley (New York: Vintage, 1978), p. 138.

22. Foucault, *The History of Sexuality, Volume 1,* pp. 138–140, emphasis mine.

Distribution
In a distributed network there are no central hubs and no satellite nodes, no trunks and no leaves. Like the rhizome, each node in a distributed network may establish direct communication with another node, without having to appeal to a hierarchical intermediary.

Charles Peterson, "Audience" (Barcelona, 1996)

The seething mass of beef cattle in Howard Hawk's *Red River* is a diagram for distribution. Ten thousand head of cattle are too large for the film and are never shown together in a single shot. Instead they appear in parts, as during the stampede, or in the end of the film when they flow down main street, heads bobbing like whitecaps on the ocean. This is what Deleuze and Guattari call a smooth space.

Deleuze recognized this, that the very site of Foucault's biopower was also a site of resistance.

Lest readers overlook its importance, he repeats his realization three times consecutively in an important section of his book *Foucault:* "[1] When power . . . takes life as its aim or object, then resistance to power already puts itself on the side of life, and turns life against power. . . . [2] Life becomes resistance to power when power takes life as its object. . . . [3] When power becomes bio-power resistance becomes the power of life, a vital power that cannot be confined within species, environment or the paths of a particular diagram."[23] Is *life resistance* a way of engaging with distributed forms of protocological management?

Part III of this book, "Protocol Futures," answers yes. While the new networked technologies have forced an ever more reticent public to adapt to the control structures of global capital, there has emerged a new set of social practices that inflects or otherwise diverts these protocological flows toward the goal of a utopian form of unalienated social life.

What is wrong with protocol? To steal a line from Foucault, it's not that protocol is bad but that protocol is *dangerous.* To refuse protocol, then, is not so much to reject today's technologies as did Theodore Kaczynski (the Unabomber), but to direct these protocological technologies, whose distributed structure is empowering indeed, toward what Hans Magnus Enzensberger calls an "emancipated media" created by active social actors rather than passive users.[24]

As Deleuze remarked to Antonio Negri several years ago:

It's true that, even before control societies are fully in place, forms of delinquency or resistance (two different things) are also appearing. Computer piracy and viruses, for example, will replace strikes and what the nineteenth century called "sabotage" . . . You ask whether control or communication societies will lead to forms of resistance

---

23. Deleuze, *Foucault,* p. 92.

24. Natalie Jeremijenko uses the rubric of "structures of participation" to think about how certain implementations of technology promote active user involvement and understanding while other technologies obfuscate understanding and control user involvement.

that might reopen the way for a communism . . . The key thing may be to create vac-
uoles of noncommunication, circuit breakers, so we can elude control.[25]

The key here is less the eluding or the breaking or the *non*communication,
but simply that Deleuze had the foresight to situate resistive action *within
the protocological field.* In the same way that biopower is a species-level knowledge, protocol is a type of species-knowledge for coded life forms. Each new
diagram, each new technology, each new management style both is an improvement on the previous one and contains with it a germ that must grow
into a still higher form. I am not suggesting that one should learn to love the
various apparatuses of control, but rather that, for all its faults, protocological control is still an improvement over other modes of social control. I hope
to show in this book that it is *through* protocol that one must guide one's efforts, not against it.

"No more vapor theory anymore," wrote Geert Lovink. Vapor theory
tends to ignore the computer itself. The computer is often eclipsed by that
more familiar thing, information society. Mine is not a book about information society, but about the real machines that live within that society.

Thus, my study skips direct engagement with the work of Alvin Toffler,
Peter Drucker, Daniel Bell, and others who discuss the third phase of capitalist development in social terms.

The large mass of literature devoted to artificial intelligence and speculations about the consciousness (or lack thereof) within man and machine
is also largely avoided in this book. Writers like Ray Kurzweil forecast a
utopian superfuture dominated by immortal man-machine hybrids. Hans
Moravec predicts a similar future, only one less populated by humans who
are said to "retire" to the mercy of their ascendant computerized progeny.

Marvin Minsky, Daniel Dennett, John Searle, Hubert Dreyfus, and others have also wrestled with the topic of artificial intelligence. But they are
not addressed here. I draw a critical distinction between this body of work,
which is concerned largely with epistemology and cognitive science, and the
critical media theory that inspires this book. Where they are concerned with

---

25. Gilles Deleuze, "Control and Becoming," in *Negotiations,* trans. Martin Joughin (New
York: Columbia University Press, 1990), p. 175.

minds and questions epistemological, I am largely concerned with bodies and the material stratum of computer technology.

My study also ignores the large mass of popular responses to the new technological age, such as Nicholas Negroponte's *Being Digital,* whose gee-whiz descriptions of the incredible *newness* of new technologies seem already dated and thin.

Except for chapter 4, this is largely *not* a book about issues specifically relating to law, Internet governance, state sovereignty, commercial power, or the like. Several books already do an excellent job covering these issues including Milton Mueller's *Ruling the Root.*

While my ultimate indebtedness to many of these authors will be obvious, it is not my goal to examine the social or culturo-historical characteristics of informatization, artificial intelligence, or virtual anything, but rather to study computers as André Bazin studied film or Roland Barthes studied the striptease: to look at a material technology and analyze its specific formal functions and dysfunctions.

To that end this book focuses on distributed computer networks and the protocological system of control present within them. I hope to build on texts such as Friedrich Kittler's groundbreaking *Discourse Networks, 1800/1900,* which describes the paradigm shift from a discourse driven by meaning and sense, to our present milieu of pattern and code. Kittler's two ages, symbolized by the two years 1800 and 1900, correspond structurally (but less so chronologically) to the social periodization supplied by Foucault and Deleuze. The passage from the modern disciplinary societies to those of the control societies, as I have already suggested, is the single most important historical transformation in this book.

Norbert Wiener is also an important character. His books laid important groundwork for how control works within physical bodies. The provocative but tantalizingly thin *Pandemonium: The Rise of Predatory Locales in the Postwar World* from architect Branden Hookway, looks at how cybernetic bodies permeate twentieth-century life. Other important theorists from the field of computer and media studies who have influenced me include Vannevar Bush, Hans Magnus Enzensberger, Marshall McLuhan, Lewis Mumford, and Alan Turing.

I am also inspired by Lovink's new school of media theory known as Net criticism. This loose international grouping of critics and practitioners has grown up with the Internet and includes the pioneering work of Hakim Bey and Critical Art Ensemble, as well as newer material from Timothy Druckrey, Marina Gržinić, Lev Manovich, Sadie Plant, and many others. Much of this intellectual work has taken place in online venues such as *CTHEORY, Nettime,* and *Rhizome,* plus conferences such as the annual Ars Electronica festival and the Next 5 Minutes series on tactical media.

Although my book is heavily influenced by film and video theory, I include here little discussion of media formats prior to the digital computer.[26] I gain much of my momentum by relying on the specificity of the digital computer as a medium, not its similarity to other visual media. In my estimation, it makes little sense to try to fit non-protocological and nondistributed media such as film and video into this new context—in the same way that it makes little sense to speak of the aura of a Web page, or the essence of a digital text. Nevertheless the history of avant-garde artistic production, from modernist painting to conceptual art, significantly influences my perspective vis-à-vis work being done today.

While lay readers may group all literature dealing with new technologies under the general heading informatization, there is an alternate path that I attempt to follow in this book. This alternate path recognizes the material substrate of media, and the historical processes that alter and create it. It attempts to chart what Manuel DeLanda calls "institutional ecologies." He writes here of the history of warfare, but it could easily refer to digital computing:

I would like to repeat my call for more realistic models of economic history, models involving the full complexity of the institutional ecologies involved, including markets, anti-markets, military and bureaucratic institutions, and if we are to believe Michel Foucault, schools, hospitals, prisons, and many others. It is only through an honest philosophical confrontation with our complex past that we can expect to understand it and derive the lessons we may use when intervening in the present and speculating about the future.[27]

---

26. For an anthology of recent writing on the confluence of cinematic practices and new media, see Martin Rieser and Andrea Zapp, eds., *New Screen Media: Cinema/Art/Narrative* (London: BFI, 2002).

27. Manuel DeLanda, "Economics, Computers, and the War Machine," in *Ars Electronica: Facing the Future,* ed. Timothy Druckrey (Cambridge: MIT Press, 1999), p. 325.

The complex "institutional ecology" of modern computing is thus the focus of this book.

Just as Marx descended into the internal structure of the commodity to interpret its material workings within the context of production at large, I must descend instead into the distributed networks, the programming languages, the computer protocols, and other digital technologies that have transformed twenty-first-century production into a vital mass of immaterial flows and instantaneous transactions.

Indeed, I attempt to read the never-ending stream of computer code *as one reads any text* (the former having yet to achieve recognition as a natural language), decoding its structure of control as one would a film or novel.

Periodization

Let me pause for a minute to address something that is taken for granted throughout much of the rest of this book. I refer to the axiom, taken from periodization theory, that history may be divided into certain broad phases, and that the late twentieth century is part of a certain phase that (although it goes by several different names) I refer to alternatively as the postmodern or digital age.

It is no mystery to scholars of critical theory that, while terminology and timelines may differ, a whole series of thinkers have roughly agreed on three broad historical phases, these being the classical era, the modern era, and the postmodern era.[28] This general consensus is what I would like to describe briefly now, not to fetishize its overarching structure, but instead to observe that "periodization *is an initial technique that opens the path and allows us to gain access to history and historical differences.*"[29] While this comparativist approach to periodization theory will undoubtedly land me in somewhat treacherous waters (for who is able to align so many different thinkers chronologically,

---

28. This triad refers primarily to the sociopolitical realm. In the realm of culture, a different triad becomes more important, that of realism, modernism, and postmodernism. See especially Fredric Jameson's extended essay "The Existence of Italy," in *Signatures of the Visible* (New York: Routledge, 1992).

29. This citation is from Michael Hardt and Kathi Weeks's interpretation of Fredric Jameson in *The Jameson Reader* (Oxford: Blackwell, 2000), p. 13, emphasis in original.

much less structurally!), I feel that the overwhelming consensus among many of my theoretical sources must be brought into the light of day before I continue with my own observation—that protocol is a system of management *historically posterior* to decentralization.

Foucault—both in his own writings, and as he has been interpreted by Deleuze—has put forth perhaps the clearest periodization. Foucault was especially interested in the historical shift from what he called the sovereign, or "classical," era of the eighteenth century, and the disciplinary, or "modern," era beginning after the French Revolution and extending into the early part of the twentieth century.

In his persuasive introduction to *Discipline and Punish,* Foucault observes that this historical transformation transpired, at least in the prison system and other systems of socialized punishment, between the years 1750 and 1830. While physical punishment was more dominant during the eighteenth century, "[a]t the beginning of the nineteenth century," writes Foucault, "the great spectacle of physical punishment disappeared . . . The age of sobriety in punishment had begun."[30] At the same time that punishment became more "sober" it also became more diffuse, more immanent to the personal lives and habits of people. Good citizens were now expected to punish *themselves,* to preemptively discipline their own bodies such that the power of punishment originated ultimately from within, not from some outside force.

This historical shift, from sovereign society to disciplinary society, reoccurs throughout the writings of Foucault, particularly in texts such as *Madness and Civilization* and *The History of Sexuality, Volume 1.* One may make the analogy that this transformation is the same as the shift from a centralized diagram (one overseer) to a decentralized diagram (many overseers).

Deleuze reinforces the historical arguments, first presented by Foucault, in his book *Foucault,* as well as in several interviews and incidental texts in the collection *Negotiations.* Deleuze's contribution was to flesh out the later segment of Foucault's periodization, and to suggest that Foucault was as clearly in tune with the second shift from disciplinarity to control as he was

---

30. Michel Foucault, *Discipline and Punish,* trans. Alan Sheridan (New York: Vintage, 1995), p. 14.

with the first shift from sovereignty to disciplinarity. While Deleuze's writings on Foucault may in fact tell readers more about Deleuze's predilections than Foucault's, nevertheless Deleuze has much to contribute, especially by establishing a connection between control society and computers (a word hardly mentioned in Foucault, if at all).

Deleuze defines the relationship between the different social phases and their native machinic technologies very clearly, in two different texts. The first comes from his 1990 interview with Antonio Negri, where he writes: "Each kind of society corresponds to a particular kind of machine—with simple mechanical machines corresponding to sovereign societies, thermodynamic machines to disciplinary societies, cybernetic machines and computers to control societies."[31] A few months later, in his "Postscript on Control Societies," Deleuze says much the same thing: "It's easy to set up a correspondence between any society and some kind of machine . . . The old sovereign societies worked with simple machines, levers, pulleys, clocks; but recent disciplinary societies were equipped with thermodynamic machines . . . ; control societies function with a third generation of machines, with information technology and computers."[32] In Deleuze, therefore, computers are historically concurrent with control societies.

Kittler agrees roughly with this periodization in his book *Discourse Networks, 1800/1900*. Reminiscent of Foucault's genealogies, Kittler's book is a history of knowledge over the last two hundred years. Kittler looks at two years—1800 and 1900—and shows how the state of knowledge changed from a "kingdom of sense" (in 1800) based on understanding and meaning to a "kingdom of pattern" (in 1900) based on images and algorithms.

He defines a discourse network as "the network of technologies and institutions that allow a given culture to select, store, and process relevant data."[33] Discourse networks change, as disciplinary networks changed for Foucault, and it is this transformation that so interests Kittler. He writes:

---

31. Deleuze, *Negotiations,* p. 175.

32. Deleuze, *Negotiations,* p. 180.

33. Friedrich Kittler, *Discourse Networks, 1800/1900,* trans. Michael Metteer and Chris Cullens (Stanford: Stanford University Press, 1990), p. 369.

In the discourse network of 1900, discourse is produced by RANDOM GENERA-TORS. Psychophysics constructed such sources of noise; the new technological media stored their output . . . The discourse network of 1900 was the first to establish a treasury of the signifier whose rules were entirely based on randomness and combinatorics . . . The discourse network of 1800 played the game of not being a discourse network and pretended instead to be the inwardness and voice of Man; in 1900 a type of writing assumes power that does not conform to traditional writing systems but rather radicalizes the technology of writing in general.[34]

Kittler's 1800 kingdom of sense corresponds roughly to Foucault's sovereign societies: Both are interested in depth, in probing to the heart of a body or an object to derive its essential meaning. 1800 is the year of the signifier.

At the same time Kittler's 1900 kingdom of pattern corresponds roughly to Foucault's disciplinary societies: Both are interested in the patterned affection of bodies and information. In what Kittler calls the "logic of chaos and intervals,"[35] the machinic processes embodied in the patterning apparatus of the typewriter or the phonograph come to the fore. 1900 is the year of the algorithm. Again, one may make the analogy that this transformation is the transformation from centralization (singular meaning) to decentralization (meaning's replication).

In the sociopolitical realm many thinkers have also charted this same periodization. Ernst Mandel uses the concept of Kondratieff waves to examine what he calls the era of late capitalism beginning in approximately 1945. "As far as I can see," writes Fredric Jameson, "the general use of the term *late capitalism* originated with the Frankfurt School; it is everywhere in Adorno and Horkheimer, sometimes varied with their own synonyms (for example, 'administered society')."[36] Jameson states that the concept is ultimately Mandel's: "There have been three fundamental moments in capitalism, each one marking a dialectical expansion over the previous stage. These are market

34. Kittler, *Discourse Networks, 1800/1900,* pp. 206, 210, 211–212.

35. Kittler, *Discourse Networks, 1800/1900,* p. 192.

36. Fredric Jameson, *Postmodernism, or, The Cultural Logic of Late Capitalism* (Durham: Duke University Press, 1991), p. xviii.

capitalism, the monopoly stage or the stage of imperialism, and our own, wrongly called postindustrial, but what might be better termed multinational capital,"[37] or to use Mandel's terminology, late capitalism.

Like other social critics of late-twentieth-century life, Jameson looks to the economic crisis of 1973 as a turning point, a moment that "somehow crystallized"[38] these new currents of postmodernity. Jameson admits that Mandel's work "is what made [his] own thoughts on 'postmodernism' possible."[39]

Sociologist Manuel Castells has also documented this transformation out of decentralization into new distributed, flexible economies in his three-volume treatise *The Information Age: Economy, Society and Culture.* Using the term "network society" (rather than Deleuze's "society of control" or Jameson's "late capitalism"), Castells shows with extensive quantitative documentation that today's sociopolitical space is dominated not by robust national economies and core industrial sectors but by "interactive networks" and "flexible accumulation."

Charting the same periodization that I rely on in this book, Castells shows how, for example, corporate business structures have changed in the last several decades from a decentralized "vertical" corporatism to a more distributed "horizontal" meshwork: "The corporation itself has changed its organizational model, to adapt to the conditions of unpredictability ushered in by rapid economic and technological change. The main shift can be characterized as the shift from vertical bureaucracies to the horizontal corporation."[40] This transformation echoes the structural difference that Deleuze and Guattari see between the tree and the rhizome.[41] Trees correspond to vertical bureaucracies, while rhizomes correspond to horizontal meshworks.

While Michael Hardt and Antonio Negri have an almost identical analysis of contemporary economics in their book *Empire,* their analysis of poli-

---

37. Jameson, *Postmodernism,* p. 35.

38. Jameson, *Postmodernism,* p. xx.

39. Jameson, *Postmodernism,* p. 400.

40. Manuel Castells, *The Information Age: Economy, Society and Culture: Volume 1—The Rise of the Network Society* (Oxford: Blackwell, 1996), p. 164, emphasis removed from original.

41. See Gilles Deleuze and Félix Guattari, *A Thousand Plateaus,* trans. Brian Massumi (Minneapolis: University of Minnesota Press, 1987), chapter 1.

tics is more sophisticated. Conscious of their relationship to Foucault and Deleuze's argument described earlier, Hardt and Negri connect the society of control to the new world order they call "Empire."

First, they define the pre-imperial forces of the disciplinary society: "[d]isciplinary society is that society in which social command is constructed through a diffuse network of *dispositifs* or apparatuses that produce and regulate customs, habits, and productive practices."[42] Then, they define the society of control as that society "in which mechanisms of command become ever more 'democratic,' ever more immanent to the social field, *distributed* throughout the brains and bodies of the citizens."[43]

Hardt and Negri specifically address new media in *Empire,* writing that, within the Internet, "[a]n indeterminate and potentially unlimited number of interconnected nodes communicate with no central point of control."[44] In their opinion this "decentralized" architecture is "what makes control of the network so difficult."[45]

While I spend much of this book arguing against such descriptions of the Internet (i.e., I argue that the Internet is distributed not decentralized and that it is in fact highly controlled despite having few if any *central* points of control), this appears to be a nonfatal mistake in their argument. The attentive reader will notice that here Hardt and Negri actually mean *modern* control and not imperial control. For what they say elsewhere about Empire should also be true here about new media. A distributed architecture is precisely that which makes protocological/imperial control of the network so easy. In fact, the various Internet protocols mandate that control may *only* be derived from such a distributed architecture.

---

42. Michael Hardt and Antonio Negri, *Empire* (Cambridge: Harvard University Press, 2000), p. 23. These "dispositifs" allude to the apparatuses of the prison or the hospital observed by Foucault, or even more specifically to the "ideological state apparatuses" and "repressive state apparatuses" observed by Foucault's teacher Louis Althusser, through whose work the term "apparatus" gained prominence in film studies and other critical theory of the late 1960s and 1970s.

43. Hardt and Negri, *Empire,* p. 23, emphasis mine.

44. Hardt and Negri, *Empire,* p. 299.

45. Hardt and Negri, *Empire,* p. 299.

Hardt and Negri confirm this position by writing elsewhere that "the passage to the society of control does not in any way mean the end of discipline [i.e., control]. In fact, the immanent exercise of discipline . . . is extended even more generally in the society of control."[46]

The computer protocol is thus in lockstep with Hardt and Negri's analysis of Empire's logics, particularly the third mode of imperial command, the *managerial* economy of command.[47] This command protocol knows from the start that "[c]ontingency, mobility, and flexibility are Empire's real power."[48] Protocological control mirrors the movements of Empire. In fact one might go so far as to say that *Empire is the social theory and protocol the technical.* Thus Hardt and Negri are accurate in their analysis of the "Symptoms of Passage." An analysis of computer protocols proves this, for it reassigns the former weapons of Leftists—celebration of difference, attack on essentialism, and so forth—as the new tools of Empire: "This new enemy not only is resistant to the old weapons but actually thrives on them, and thus joins its would-be antagonists in applying them to the fullest. Long live difference! Down with essentialist binaries."[49] A distributed network is precisely what gives IP its effectiveness as a dominant protocol. Or to take another example, the flimsy, cross-platform nature of HTML is precisely what gives it its power as a protocological standard. Like Empire, if protocol dared to centralize, or dared to hierarchize, or dared to essentialize, it would fail.

Further to these many theoretical interventions—Foucault, Deleuze, Kittler, Mandel, Castells, Jameson, Hardt and Negri—are many dates that roughly confirm my periodization: the discovery of DNA in 1953; the economic crisis in the West during the 1970s epitomized by President Richard Nixon's decoupling of the U.S. dollar from the gold standard on August 17, 1971 (and thus the symbolic evaporation of the Bretton Woods agreement); Charles Jencks's claim that modern architecture ended on July 15, 1972, at 3:32 P.M.; the ARPAnet's mandatory rollover to TCP/IP on January 1, 1983; the fall of the Berlin Wall in 1989; the crashing of AT&T's long-distance

---

46. Hardt and Negri, *Empire,* p. 330.
47. Hardt and Negri, *Empire,* p. 199.
48. Hardt and Negri, *Empire,* p. 200.
49. Hardt and Negri, *Empire,* p. 138.

Table 1

Periodization Map

Period	Machine	Dates	Diagram	Manager
Sovereign society	Simple mechanical machines	March 2, 1757 (Foucault)	Centralization	Hierarchy
Disciplinary society	Thermodynamic machines	May 24, 1844 (telegraph); 1942 (Manhattan Project)	Decentralization	Bureaucracy
Control society	Cybernetic machines, computers	February 28, 1953 (Watson and Crick); January 1, 1983 (TCP/IP)	Distribution	Protocol

telephone switches on January 15, 1990; the start of the Gulf War on January 17, 1991.[50] These dates, plus the many periodization theories mentioned earlier, map together as shown in table 1.

That these dates do not line up in any precise manner is of no concern. Periodization theory is a loose art at best and must take into account that, when history changes, it changes slowly and in an overlapping, multilayered way, such that one historical moment may extend well into another, or two moments may happily coexist for decades or longer. For instance, in much of the last hundred years, *all three social phases described earlier existed at the same time* in the United States and elsewhere. To paraphrase William Gibson: The future is already here, but it is not uniformly distributed across all points in society. At best, periodization theory is an analytical mindgame, yet one that breathes life into the structural analyses offered to explain certain tectonic shifts in the foundations of social and political life. My book implicitly participates in this game, mapping out certain details of the third, "control society" phase, specifically the diagram of the distributed network, the technology of the computer, and the management style of protocol.

---

50. For Jencks, see his *The Language of Post-Modern Architecture* (New York: Rizzoli, 1991); for references to AT&T, see Bruce Sterling's *The Hacker Crackdown* (New York: Bantam, 1993) and Michelle Slatalla and Joshua Quittner's *Masters of Deception* (New York: HarperCollins, 1995).

1

Physical Media

The language of the RFC was warm and welcoming.
—KATIE HAFNER AND MATTHEW LYON, *Where Wizards Stay Up Late*

While many have debated the origins of the Internet, it's clear that in many ways it was built to withstand nuclear attack. The Net was designed as a solution to the vulnerability of the military's centralized system of command and control during the late 1950s and beyond. For, the argument goes, if there are no central command centers, then there can be no central targets and overall damage is reduced.

If one can consider nuclear attack as the most highly energetic, dominating, and centralized force that one knows—an archetype of the modern era—then the Net is at once the solution to and inversion of this massive material threat, for it is precisely noncentralized, nondominating, and nonhostile.

The term *protocol* is most known today in its military context, as a method of correct behavior under a given chain of command. On the Internet, the meaning of protocol is slightly different. In fact, the reason why the Internet would withstand nuclear attack is precisely because its internal protocols are the enemy of bureaucracy, of rigid hierarchy, and of centralization. As I show in this chapter, the material substrate of network protocols is highly flexible, distributed, and resistive of hierarchy.

The packet-switching technologies behind the Internet provided a very different "solution" to nuclear attack than did common military protocol during the Cold War. For example, in 1958 the Royal Canadian Air Force and the U.S. Air Force entered into agreement under the North American Aerospace Defense Command (NORAD). NORAD is a radar surveillance system ringing North America that provides early warnings of missile or other air attacks against Canada and the United States. "The command monitors any potential aerospace threat to the two nations, provides warning and assessment of that threat for the two governments, and responds defensively to any aircraft or cruise missile threatening North American airspace."[1] The NORAD system is a centralized, hierarchical network. It contains regional control sectors, all of which are ultimately controlled by the USSPACECOM Command Center at Cheyenne Mountain in Colorado Springs, Colorado. It functions like a wall, not like a meshwork. Faced with a nuclear attack,

Epigraph: Katie Hafner and Matthew Lyon, *Where Wizards Stay Up Late: The Origins of the Internet* (New York: Touchstone, 1996), p. 144.

1. *NORAD: Into the 21st Century,* U.S. Government Printing Office (1997-574-974).

NORAD meets force with force. Once the outer protection zone of the land-mass is compromised, the NORAD command is able to scramble defensive air power through a rigidly defined system of command and control that is directed outward from a single source (USSPACECOM), to subservient end-point installations that help resist attack. The specific location of each radar installation is crucial, as is the path of the chain of command. During the Cold War, NORAD was the lynchpin of nuclear defense in North America. It is a "solution" to the nuclear threat.

The Internet system could not be more different. It follows a contrary organizational design. The Internet is based not on directionality nor on toughness, but on flexibility and adaptability. Normal military protocol serves to hierarchize, to prioritize, while the newer network protocols of the Internet serve to *distribute.*

In this chapter I describe exactly what distribution means, and how protocol works in this new terrain of the distributed network.[2] I attempt to show that protocol is not by nature horizontal or vertical, but that protocol is an algorithm, *a proscription for structure* whose form of appearance may be any number of different diagrams or shapes.

The simplest network diagram is the centralized network (see figure 1.1). Centralized networks are hierarchical. They operate with a single authoritative hub. Each radial node, or branch of the hierarchy, is subordinate to the central hub. All activity travels from center to periphery. No peripheral node is connected to any other node. Centralized networks may have more than one branch extending out from the center, but at each level of the hierarchy power is wielded by the top over the bottom.

---

2. The division of network designs into centralized, decentralized, and distributed appears in Paul Baran's *On Distributed Communications: 1. Introduction to Distributed Communications Networks* (Santa Monica, CA: RAND, 1964), p. 2. Baran's diagrams have been copied by many authors since then.

Following William Evan, John Arquilla and David Ronfeldt suggest a topology even simpler than the centralized network. This is the chain or line network: for example, "in a smuggling chain where people, goods, or information move along a line of separate contacts, and where end-to-end communication must travel through the intermediate nodes." See Arquilla and Ronfeldt, *Networks and Netwars: The Future of Terror, Crime, and Militancy* (Santa Monica, CA: RAND, 2001), p. 7.

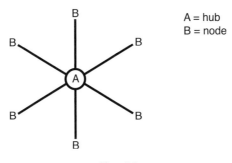

A = hub
B = node

Figure 1.1
A centralized network

The American judicial system, for example, is a centralized network. While there are many levels to the court system, each with its own jurisdiction, each decision of each court can always be escalated (through the appeals process) to a higher level in the hierarchy. Ultimately, however, the Supreme Court has final say over all matters of law.

The panopticon, described in Foucault's *Discipline and Punish,* is also a centralized network. In the panopticon, repurposed by Foucault from the writings of Jeremy Bentham, a guard is situated at the center of many radial cells. Each cell contains a prisoner. This special relationship between guard and prisoner "links the centre and periphery." In it, "power is exercised without division, according to a continuous hierarchical figure" occupying the central hub.[3]

A *de*centralized network is a multiplication of the centralized network (see figure 1.2). In a decentralized network, instead of one hub there are many hubs, each with its own array of dependent nodes. While several hubs exist, each with its own domain, no single zenith point exercises control over all others.

There are many decentralized networks in the world today—in fact, decentralized networks *are the most common diagram of the modern era.*

One example is the airline system. In it, one must always travel through certain centralized hub cities—generally in the Midwest or central areas of the United States. Direct nonstop service is only possible if one happens to

---

3. Michel Foucault, *Discipline and Punish,* trans. Alan Sheridan (New York: Vintage, 1997), p. 197.

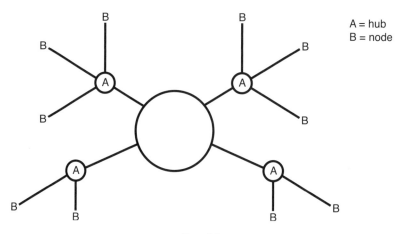

Figure 1.2
A decentralized network

be traveling from one hub to another (or if one pays a premium for special routes).

For the airline system, the decentralized network is the solution to multiplicity, albeit a compromise between the needs of the passenger and the needs of the airlines. There are far too many airports in the country to allow for nonstop service between each and every city; however, it would be inefficient to route every passenger through a single, Midwestern hub (e.g., consider a flight from North Carolina to Maine).

The third network diagram, the one that interests me most here, is called the distributed network.[4] The emergence of distributed networks is part of a larger shift in social life. The shift includes a movement away from central

4. In *Networks and Netwars,* Arquilla and Ronfeldt call this third network topology an "all-channel" network "where everybody is connected to everybody else" (p. 8). However their all-channel network is not identical to a distributed network, as their senatorial example betrays: "an all-channel council or directorate" (p. 8). Truly distributed networks cannot, in fact, support all-channel communication (a combinatorial utopia), but instead propagate through outages and uptimes alike, through miles of dark fiber (Lovink) and data oases, through hyperskilled capital and unskilled laity. Thus distribution is similar to but not synonymous with all-channel, the latter being a mathematical fantasy of the former.

bureaucracies and vertical hierarchies toward a broad network of autonomous social actors.

As Branden Hookway writes: "The shift is occurring across the spectrum of information technologies as we move from models of the global application of intelligence, with their universality and frictionless dispersal, to one of local applications, where intelligence is site-specific and fluid."[5] Computer scientists reference this historical shift when they describe the change from linear programming to *object-oriented* programming, the latter a less centralized and more modular way of writing code. This shift toward distribution has also been documented in such diverse texts as those of sociologist Manuel Castells, American Deleuzian Hakim Bey, and the Italian "autonomist" political movement of the 1970s. Even harsh critics of this shift, such as Nick Dyer-Witheford, surely admit that the shift is taking place. It is part of a larger process of postmodernization that is happening the world over.

What is the nature of these distributed networks? First, distributed networks have no central hubs and no radial nodes. Instead each entity in the distributed network is an autonomous agent.

A perfect example of a distributed network is the rhizome described in Deleuze and Guattari's *A Thousand Plateaus.* Reacting specifically to what they see as the totalitarianism inherent in centralized and even decentralized networks, Deleuze and Guattari instead describe the rhizome, a horizontal meshwork derived from botany. The rhizome links many autonomous nodes together in a manner that is neither linear nor hierarchical. Rhizomes are heterogeneous and connective, that is to say, "any point of a rhizome can be connected to anything other."[6] They are also multiple and asymmetrical: "[a] rhizome may be broken, shattered at a given spot, but it will start up again on one of its old lines, or on new lines."[7] Further, the rhizome has complete disregard for depth models, or procedures of derivation. As Deleuze and Guattari write, a rhizome "is a stranger to any idea of genetic axis

---

5. Branden Hookway, *Pandemonium: The Rise of Predatory Locales in the Postwar World* (New York: Princeton Architectural Press, 1999), pp. 23–24.

6. Gilles Deleuze and Félix Guattari, *A Thousand Plateaus,* trans. Brain Massumi (Minneapolis: University of Minnesota Press, 1987), p. 7.

7. Deleuze and Guattari, *A Thousand Plateaus,* p. 9.

or deep structure."[8] Trees and roots, and indeed "[a]ll of arborescent culture"[9] is rejected by the rhizome. Summarizing the unique characteristics of the rhizome—and with it the distributed network—Deleuze and Guattari write:

• [U]nlike trees or their roots, the rhizome connects any point to any other point . . .
• The rhizome is reducible neither to the One nor the multiple. . . . It is composed not of units but of dimensions, or rather directions in motion.
• It has neither beginning nor end, but always a middle (*milieu*) from which it grows and which it overspills.
• Unlike a structure, which is defined by a set of points and positions, with binary relations between the points and biunivocal relationships between the positions, the rhizome is made only of lines . . .
• Unlike the tree, the rhizome is not the object of reproduction . . .
• The rhizome is an antigenealogy. It is short-term memory, or antimemory.
• The rhizome operates by variation, expansion, conquest, capture, offshoots.
• The rhizome is an acentered, nonhierarchical, nonsignifying system without a General and without an organizing memory or central automation.[10]

If diagrammed, a distributed network might look like figure 1.3. In a distributed network, each node *may* connect to any other node (although there is no requirement that it does). During a node-to-node connection, no intermediary hubs are required—none, not even a centralized switch as is the case in the telephone network. Point "X" may contact "Y" directly via one of several path combinations.

A distributed network is always caught, to use an expression from Deleuze and Guattari, au milieu, meaning that it is never complete, or integral to itself. The lines of a distributed network continue off the diagram. Any subsegment of a distributed network is as large and as small as its parent network. Distribution propagates through rhythm, not rebirth.

---

8. Deleuze and Guattari, *A Thousand Plateaus,* p. 12.

9. Deleuze and Guattari, *A Thousand Plateaus,* p. 15.

10. Deleuze and Guattari, *A Thousand Plateaus,* p. 21, bulleted format mine.

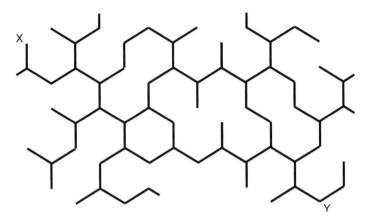

Figure 1.3
A distributed network

One actually existing distributed network is the Dwight D. Eisenhower
System of Interstate & Defense Highways, better known as the interstate
highway system. The highway system was first approved by Congress immediately following World War II, but was not officially begun until June
29, 1956, when President Eisenhower signed it into law. (This is exactly the
same period during which Internet pioneer Paul Baran began experiment-
ing with distributed, packet-switching computer technologies at the Rand
Corporation.[11]) The highway system is a distributed network because it lacks
any centralized hubs and offers direct linkages from city to city through a variety of highway combinations.

For example, someone traveling from Los Angeles to Denver may begin
by traveling on Interstate 5 north toward San Francisco turning northwest
on Interstate 80, or head out on Interstate 15 toward Las Vegas, or even Interstate 40 toward Albuquerque. The routes are varied, not predetermined.
If one route is blocked, another will do just as well. These are the advantages
of a distributed network.

---

11. As Hafner and Lyon write: "Baran was working on the problem of how to build communication structures whose surviving components could continue to function as a cohesive entity after other pieces were destroyed." See Katie Hafner and Matthew Lyon, *Where Wizards
Stay Up Late*, p. 56.

THE AT&T GLOBAL NETWORK
OPERATIONS CENTER

AT&T Global Network Operation Center (architect: HOK; photo: Peter Paige)

Centralization & Decentralization

A centralized network consists of a single central power point (a host), from which are attached radial nodes. The central point is connected to all of the satellite nodes which are themselves connected only to the central host. A decentralized network, on the other hand, has *multiple* central hosts, each with its own set of satellite nodes. A satellite node may have connectivity with one or more hosts, but not with other nodes. Communication generally travels unidirectionally within both centralized and decentralized networks: from the central trunks to the radial leaves.

Of course the Internet is another popular and actually existing distributed network. Both the Internet and the U.S. interstate highway system were developed in roughly the same time period (from the late 1950s to the late 1970s), for roughly the same reason (to facilitate mobility and communication in case of war). Later, they both matured into highly useful tools for civilians.

What was once protocol's primary liability in its former military context—the autonomous agent who does not listen to the chain of command—is now its primary constituent in the civil context. The diagram for protocol has shifted from the centralized to the decentralized network, and now finally to the distributed network. Distributed networks have no chain of command, only autonomous agents who operated according to certain pre-agreed "scientific" rules of the system.

For the Internet, these scientific rules are written down. Called protocols, they are available in documents known as RFCs, or "Requests for Comments." Each RFC acts as a blueprint for a specific protocol. It instructs potential software designers and other computer scientists how to correctly implement each protocol in the real world. Far more than mere technical documentation, however, the RFCs are a discursive treasure trove for the critical theorist.

The RFC on "Requirements for Internet Hosts," an introductory document, defines the Internet as a series of interconnected networks, that is, a *network of networks,* that are interconnected via numerous interfacing computers called gateways: "An Internet communication system consists of interconnected packet networks supporting communication among host computers using the Internet protocols . . . The networks are interconnected using packet-switching computers called 'gateways.'"[12] Populating these many different networks are hosts, single computers that are able to send and receive information over the network. According to this RFC, "A host computer, or simply 'host,' is the ultimate consumer of communication services. A host generally executes application programs on behalf of user(s), employing network and/or Internet communication services in support of this function. . . . Internet hosts span a wide range of size, speed, and function.

---

12. Robert Braden, "Requirements for Internet Hosts," RFC 1122, October 1989, p. 6.

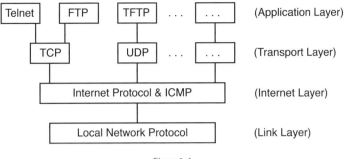

Figure 1.4
Protocol layers

They range in size from small microprocessors through workstations to main-frames and supercomputers."[13] Or, as the RFC on Transmission Control Protocol simply defines it, hosts are "computers attached to a network."[14] If the host is a receiver of information, it is called a client. If it is a sender of information, it is called a server.

In order for hosts to communicate via the Internet, they must implement an entire suite of different protocols. Protocols are the common languages that all computers on the network speak. These component protocols act like layers. Each layer has a different function (see figure 1.4). Considered as a whole, the layers allow communication to happen.

The RFC on "Requirements for Internet Hosts" defines four basic layers for the Internet suite of protocols: (1) the application layer (e.g., telnet, the Web), (2) the transport layer (e.g., TCP), (3) the Internet layer (e.g., IP), and (4) the link (or media-access) layer (e.g., Ethernet).

These layers are nested, meaning that the application layer is encapsulated within the transport layer, which is encapsulated with the Internet layer, and so on.

This diagram, minus its "layer" captions, appears in RFC 791. The four layers are part of a larger, seven-layer model called the OSI (Open Systems Interconnection) Reference Model developed by the International Organization for Standardization (ISO). Tim Berners-Lee, inventor of the Web, uses a

---

13. Braden, "Requirements for Internet Hosts," pp. 6–7.

14. Jonathan Postel, "Transmission Control Protocol," RFC 793, September 1981, p. 7.

slightly different four-layer model consisting of "the transmission medium, the computer hardware, the software, and the content." Yochai Benkler, from whom Lawrence Lessig has drawn, uses instead a three-layer model consisting of a physical layer, a code layer, and a content layer. Lev Manovich uses an even simpler, two-layer model consisting of a "cultural" layer comprised of "the encyclopedia and the short story; story and plot; composition and point of view; mimesis and catharsis; comedy and tragedy," and a "computer" layer comprised of computer languages, variables, functions, packets, and other code elements.[15]

Consider an average telephone conversation as an analogy. There are several protocols at play during a telephone call. Some are technical, some social. For example, the act of listening for a dial tone and dialing the desired phone number can be considered to be in a different "layer" than the conversation itself.

Furthermore, the perfunctory statements that open and close a telephone conversation—"Hello," "Hi, this is . . . ," "Well, I'll talk to you later," "Okay, goodbye," "Bye!"—are themselves not part of the normal conversation "layer" but are merely necessary to establish the beginning and end of the conversation.

The Internet works the same way. The application layer is like the conversation layer of the telephone call. It is responsible for the content of the specific technology in question, be it checking one's email, or accessing a Web page. The application layer is a *semantic* layer, meaning that it is responsible for preserving the content of data within the network transaction. The application layer has no concern for larger problems such as establishing net-

---

15. For these references, see Jonathan Postel, "Internet Protocol," RFC 791, September 1981, p. 5; Tim Berners-Lee, *Weaving the Web* (New York: HarperCollins, 1999), pp. 129–130; Yochai Benkler's "From Consumers to Users: Shifting the Deeper Structures of Regulation Toward Sustainable Commons and User Access," *Federal Communications Law Journal* 52 (2000), pp. 561–579; and Lev Manovich, *The Language of New Media* (Cambridge: MIT Press, 2001), p. 46. The critical distinction is that the OSI model, my preferred heuristic, considers everything to be code and makes no allowances for special anthropomorphic uses of data. This makes it much easier to think about protocol. The other models privilege human-legible forms, whose reducibility to protocol is flimsy at best.

work connections, or actually sending data between those connections. It simply wants its "conversation" to work correctly.

The transport layer is one step higher in the hierarchy than the application layer. It has no concern for the content of information (one's email, one's Web page). Instead, the transport layer is responsible for making sure that the data traveling across the network arrives at its destination correctly. It is a social layer, meaning that it sits halfway between the content or meaning of the data being transferred and the raw act of transferring that data. If data is lost in transit, it is the transport layer's responsibility to resend the lost data.

Thus, in our hypothetical telephone conversation, if one hears static on the line, one might interject the comment, "Hello . . . Are you still there?" This comment is *not* part of the conversation layer (unless your conversation happens to be about "still being there"); rather, it is an interstitial comment meant to confirm that the conversation is traveling correctly across the telephone line. The opener and closer comments are also part of the transport layer. They confirm that the call has been established and that it is ready for the conversation layer, and conversely that the conversation is finished and the call will be completed.

The third layer is the Internet layer. This layer is larger still than both the application and transport layers. The Internet layer is concerned with one thing: the actual movement of data from one place to another. It has no interest in the content of that data (the application layer's responsibility) or whether parts of the data are lost in transit (the transport layer's responsibility).

The fourth layer, the link layer, is less important to my study. It is the hardware-specific layer that must ultimately encapsulate any data transfer. Link layers are highly variable due to the many differences in hardware and other physical media. For example, a telephone conversation can travel just as easily over normal telephone wire as it can over fiber-optic cable. However, in each case the technology in question is radically different. These technology-specific protocols are the concern of the link (or media-access) layer.

The different responsibilities of the different protocol layers allow the Internet to work effectively. For example, the division of labor between the transport layer and the Internet layer, whereby error correction is the sole responsibility of the transport layer and routing (the process by which data is "routed," or sent toward its final destination) is the sole responsibility of the Internet layer, creates the conditions of existence for the distributed network.

Thus, if a router goes down in Chicago while a message is en route from New York to Seattle, the lost data can be resent via Louisville instead (or Toronto, or Kansas City, or Lansing, or myriad other nodes). It matters not if the alternate node is smaller or larger, or is on a different subnetwork, or is in another country, or uses a different operating system.

The RFCs state this quality of flexibility very clearly:

A basic objective of the Internet design is to tolerate a wide range of network characteristics—e.g., bandwidth, delay, packet loss, packet reordering, and maximum packet size. Another objective is robustness against failure of individual networks, gateways, and hosts, using whatever bandwidth is still available. Finally, the goal is full "open system interconnection": an Internet host must be able to interoperate robustly and effectively with any other Internet host, across diverse Internet paths.[16]

As long as the hosts on the network conform to the general suite of Internet protocols—like a lingua franca for computers—then the transport and Internet layers, working in concert, will take care of everything.

The ultimate goal of the Internet protocols is totality. The virtues of the Internet are robustness, contingency, interoperability, flexibility, heterogeneity, pantheism. Accept everything, no matter what source, sender, or destination.

TCP is the most common protocol in the transport layer. It works very closely with the IP to ensure that the data sent via IP arrives correctly. TCP creates a "virtual circuit" between sender and recipient and uses that imaginary circuit to regulate the flow of information. Where IP is blind to the ultimate integrity of the data it transports (more on IP later), TCP constantly checks to see if the message arrives in one piece. As the RFC specifies, "TCP is used by those applications needing reliable, connection-oriented transport service, e.g., mail (SMTP), file transfer (FTP), and virtual terminal service (Telnet)."[17]

TCP is responsible for the "handshake" that happens between two computers at the moment a connection is established.

---

16. Braden, "Requirements for Internet Hosts," p. 8.
17. Braden, "Requirements for Internet Hosts," p. 82.

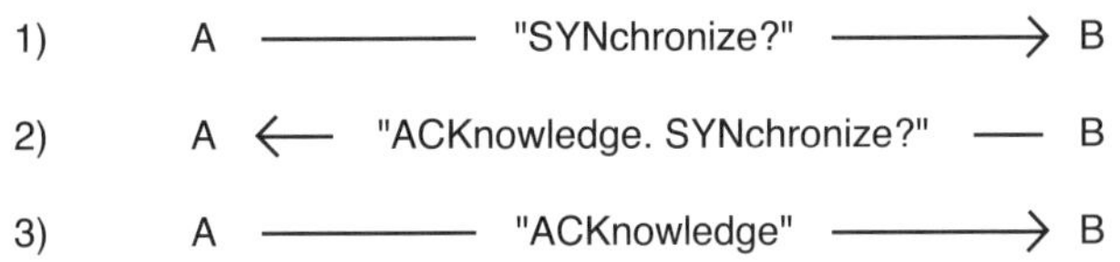

```
1)     A ————————— "SYNchronize?" ————————→ B

2)     A ⟵——— "ACKnowledge. SYNchronize?" ——— B

3)     A ————————— "ACKnowledge" ————————→ B
```

Figure 1.5
Three-way handshake

TCP creates an imaginary circuit between sender and receiver. It "saves state"; that is, it remembers the state of the conversation from moment to moment (something that IP does not do by itself, nor does the other common transport protocol called UDP). This is what the RFC refers to when it describes TCP as "a connection-oriented, end-to-end reliable protocol,"[18] as an example of ongoing "inter-process communication," or as the creation of a "logical circuit" between two computers. The circuit doesn't in fact exist in the real world, but it is created temporarily to connect sender and receiver, in much the same way that a circuit is temporarily created between caller and recipient during a normal telephone conversation (except that with the phone system, the circuit is created by an actual switch, rather than through a distributed connection).

The TCP circuit is created through a three-step process known as a handshake. First, the sender sends a message called a "SYN" (synchronize). Second, the recipient replies with a message called an "ACK" (acknowledge) and initiates its own SYN request. Finally, the original sender acknowledges the recipient's SYN by sending its own ACK (see figure 1.5). After this three-way handshake is complete—(1) "Hello!" (2) "Hi. How are you?" (3) "I'm fine thanks"—the connection is established and normal communication may begin.

The primary value of TCP is its robust quality. TCP allows communication on the Web to be very reliable: Information is monitored during transport and is re-sent if lost or corrupted.

As a system this robustness is achieved by following a general principle: "Be conservative in what you do, be liberal in what you accept from others."[19]

---

18. Postel, "Transmission Control Protocol," p. 1.

19. Postel, "Transmission Control Protocol," p. 14.

This means that TCP hosts should liberally accept as much information as possible from other, foreign devices. But if any of the information is corrupted, the "conservative" host will delete the information and request a fresh copy be re-sent. As the RFC notes, the goal of TCP is "robustness in the presence of communication unreliability and availability in the presence of congestion."[20]

TCP's partner protocol is IP. TCP and IP work together to create a protocol suite, referred to simply as TCP/IP. IP is responsible for one thing: moving small packets of data called "datagrams" from one place to another. As the RFC specifications for IP note, "the internet protocol provides for transmitting blocks of data called datagrams from sources to destinations."[21]

However, in IP there are "no mechanisms to augment end-to-end data reliability, flow control, sequencing, or other services commonly found in host-to-host protocols"[22] such as TCP. This means that IP simply seals up its datagrams and shoots them out into the ether. It does not wait for any SYNs or ACKs, and it receives no certification that the datagrams have been received (since these are all the responsibilities of the transport layer, TCP). The IP knows that, eventually, its datagrams will arrive at their locations, and if they don't, the transport layer will provide all error correction and send requests for the missing datagrams to be re-sent.

IP is like the engine powering a car—the engine moves the car, but it has no faculties for knowing when and where to steer, or knowing when and where to stop or speed up (these are the responsibilities of the driver). The engine cannot recognize the different between a green and red traffic light. It has no business dealing with things that are outside its protocological purview.

Technically, then, IP is responsible for two things: routing and fragmentation. Routing is the process by which paths are selected for moving data across a network. Since networks are heterogeneous and ever-changing, the route between point A and point B is never fixed but must be rethought each time material wishes to pass over it.

---

20. Postel, "Transmission Control Protocol," p. 1.

21. Postel, "Internet Protocol," p. 1.

22. Postel, "Internet Protocol," p. 1.

This flexible routing system is achieved through a "hopping" process whereby data is passed from computer to computer in sequence. None of the computers in the chain of hops knows definitively where the desired destination lies. But they do know in which general direction the destination is. They pass their datagrams to the computer that lies in the "general direction" of the destination. Each computer en route maintains a cache containing information about which of its neighbors lie in which general direction. Each node in the network knows not where the final destination is, but simply which direction, or "next-hop," will get it closer to its destination. If the next-hop proves to be faulty, then the intermediary gateway alerts the source computer and the source computer updates its next-hop cache.

Thus, if Chicago is the next-hop for a message leaving New York en route to Seattle, and Chicago goes down, then Louisville becomes New York's next-hop for Seattle. Later, if Chicago is reinstated and becomes the best routing option again, New York updates its cache accordingly.

The next-hop strategy means that no single node on the Internet knows definitively where a destination is, merely that it is "over there." Each node does know the exact location of every node *it is connected to,* and may pass its messages to whatever machine is closest to "over there." After enough hops in the right direction, the destination machine will no longer be "over there" but will actually be the next-hop for the router currently carrying the data, and the data will be delivered. In this way the message hops around until it arrives in the immediate vicinity of its destination, whereby the exact location of the destination is in fact known and final delivery is possible.

Each datagram is given a number called a "time-to-live." This number designates the maximum number of hops that that datagram is able to take before it is deleted. At each hop, the time-to-live is decreased by one. If the time-to-live reaches zero, the routing computer is obligated to delete the datagram. This ensures that datagrams will not hop around the network indefinitely, creating excess congestion.

The second responsibility of the Internet Protocol is fragmentation. When messages are sent across the network, they are inevitably too large to be sent in one piece. Hence, each message is fragmented, or disintegrated into several small packets, before it is sent. Each small packet is sent over the network individually. At the end, the packets are collected and reassembled to recreate the original message. This process is called fragmentation.

Each physical network has its own personalized threshold for the largest packet size it can accommodate. Thus, no single fragmentation recipe will work for every network. Some, like large freeways, will accommodate large packets, while others, like small back roads, will accommodate only small packets.

But if a message starts its journey as large packets, it cannot be stymied mid-journey if it happens to come upon a foreign network that only accommodates small packet sizes. Refragmentation may be necessary en route. Thus, if a message starts off being fragmented into large packets (e.g., if it is traveling over a fiber-optic cable), it may need to refragment itself mid-journey if it encounters a medium-sized pipe (e.g., a telephone line) somewhere en route. IP can deal with this contingency. Fragmentation allows the message to be flexible enough to fit through a wide range of networks with different thresholds for packet size.

Whenever a packet is created via fragmentation, certain precautions must be taken to make sure that it will be reassembled correctly at its destination. To this end, a header is attached to each packet. The header contains certain pieces of vital information such as its source address and destination address. A mathematical algorithm or "checksum" is also computed and amended to the header. If the destination computer determines that the information in the header is corrupted in any way (e.g., if the checksum does not correctly correlate), it is obligated to delete the packet and request that a fresh one be sent.

At this point, let me pause to summarize the distinct protocological characteristics of the TCP/IP suite:

• TCP/IP facilitates peer-to-peer communication, meaning that Internet hosts can communicate directly with each other without their communication being buffered by an intermediary hub.
• TCP/IP is a distributed technology, meaning that its structure resembles a meshwork or rhizome.
• TCP/IP is a universal language, which if spoken by two computers allows for internetworking between those computers.
• The TCP/IP suite is robust and flexible, not rigid or tough.
• The TCP/IP suite is open to a broad, theoretically unlimited variety of computers in many different locations.

- The TCP/IP protocol, and other protocols like it, is a *result* of the action of autonomous agents (computers).

Each of these characteristics alone is enough to distinguish protocol from many previous modes of social and technical organization. Together they compose a new, sophisticated system of distributed control.

Not every protocol is concerned with the process of peer-to-peer communication as are TCP and IP. DNS, or Domain Name System, is a protocol with a very simple, but different, mandate. DNS is responsible for translating Internet addresses from names to numbers.

While many computer users are familiar with the "dot-com" style of writing Internet addresses (e.g., www.superbad.com or www.rhizome.org), computers themselves use a numerical moniker instead, called an IP address. IP addresses are written as a group of four numbers separated by dots (e.g., 206.252.131.211). While it is very difficult for humans to remember and use such numbers, it is very easy for computers. "The basic problem at hand," writes DNS critic Ted Byfield, is "how we map the 'humanized' names of DNS to 'machinic' numbers of the underlying IP address system."[23] Computers understand numbers more easily, humans understand words. Thus, before each and every transaction on the World Wide Web, one's hand-typed Web address must first be translated to an IP address before the computer can do its work:

```
www.rhizome.org ↔ 206.252.131.211
```

This translation is called "resolution" and it is the reason why DNS exists. If DNS had never been developed, Internet addresses would look more like long telephone numbers or postal codes. Instead they look like long words.

Prior to the introduction of DNS in 1984, a single computer, called a *name server*, held all the name-to-number conversions. They were contained in a single text file. There was one column for all the names and another for all the numbers—like a simple reference table. This document, called HOSTS.TXT,

---

23. Ted Byfield, "DNS: A Short History and a Short Future," *Nettime,* October 13, 1998.

lived in Menlo Park, California, at the Network Information Center of the Stanford Research Institute (SRI-NIC).[24] Other computers on the Internet would consult this document periodically, downloading its information so that their local reference tables would carry the most up-to-date data. The entire system of naming referred to in this file was called the *name space.*

This early system was a centralized network, par excellence, with SRI-NIC at the center. However as the Internet grew larger this single, central node became incompatible with the nature of the network: "The toll on SRI-NIC, in terms of the network traffic and processor load involved in distributing the file, was becoming unbearable. . . . Maintaining consistency of the files across an expanding network became harder and harder. By the time a new HOSTS.TXT could reach the farthest shores of the enlarged ARPAnet, a host across the network had changed addresses, or a new host had sprung up that users wanted to reach."[25]

To solve this problem, computer scientist Paul Mockapetris designed a new system, a decentralized database of name/number mappings called DNS (see figure 1.6). The new system, still in place today, operates like an inverted tree:

The domain name space is a tree structure. Each node and leaf on the tree corresponds to a resource set (which may be empty). . . . The domain name of a node or leaf is the path from the root of the tree to the node or leaf. By convention, the labels that compose a domain name are read left to right, from the most specific (lowest) to the least specific (highest).[26]

The tree structure allows Mockapetris to divide the total name space database into more manageable and decentralized zones through a process of hierarchization. As Mockapetris writes, "approaches that attempt to collect a consistent copy of the entire database will become more and more expensive

---

24. See Paul Albitz and Cricket Liu, *DNS and BIND, Third Edition* (Sebastopol, CA: O'Reilly, 1998), p. 3.

25. Albitz and Liu, *DNS and BIND,* pp. 3–4.

26. Paul Mockapetris, "Domain Names—Concepts and Facilities," RFC 882, November 1983, p. 6.

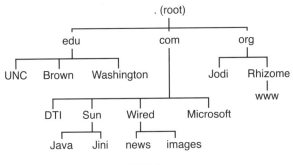

Figure 1.6
Domain Name System (DNS)

and difficult, and hence should be avoided."[27] Instead each portion of the database is delegated outward on the branches of the tree, into each leaf.

At the top of the inverted tree sit the so-called root servers, represented by a single dot (".") They have authority over the *top-level domains* (TLDs) such as "com," "net," "edu," and "org." At each branch of the tree, control over a different zone of the name space is delegated to a server that is lower on the tree. Thus, in order to resolve the address "www.rhizome.org," one must first ask the root server where to find the "org" zone. The root server replies with an authoritative answer about where to find the "org" name server. Then, the "org" name server is queried and replies with the answer for where to find the "rhizome" host within the "org" zone. Finally, the "rhizome" name server is queried, and replies with the numerical address for the "www" computer that lives within the "rhizome" domain.

Like this, the process starts at the most general point, then follows the chain of delegated authority until the end of the line is reached and the numerical address may be obtained. This is the protocol of a decentralized network.

In DNS, each name server can reply only with authoritative information about the zone immediately below it. This is why the system is hierarchical. But each name server can *only* know authoritative information about the zone immediately below it. The second, or third, or even fourth segment down the branch has been delegated to other name servers. This is why the system is decentralized.

---

27. Mockapetris, "Domain Names—Concepts and Facilities," p. 2.

The more central name servers that are closer to the root of the tree cannot tell you authoritative information about the computers at the ends of the branches, but they *can* tell you who they have delegated such information to and where to find the delegates.

As mentioned in the introduction to this book, protocol is based on a contradiction between two opposing machinic technologies: One radically distributes control into autonomous locales (exemplified here by TCP and IP), and the other focuses control into rigidly defined hierarchies (exemplified here by DNS). There are other important conclusions that one may derive from the preceding discussion of protocol.

First, as the discussion of DNS suggests, protocol is a universalizing system. Ted Byfield writes that what is unique to the DNS is

> its historical position as the first "universal" addressing system—that is, a naming convention called upon . . . to integrate not just geographical references at every scale . . . but also commercial language of every type (company names, trademarks, jingles, acronyms, services, commodities), proper names (groups, individuals), historical references (famous battles, movements, books, songs), hobbies and interests, categories and standards (concepts, specifications, proposals) . . . the list goes on and on.[28]

DNS is the most heroic of human projects; it is the actual construction of a single, exhaustive index for all things. It is the encyclopedia of mankind, a map that has a one-to-one relationship with its territory. Thus, as I demonstrate in chapter 2, DNS is like many other protocols in that, in its mad dash toward universality, it produces sameness or consistency where originally there existed arbitrariness. As the saying goes, apples and oranges are not comparable in the "real world," but in the DNS system they are separated by a few binary digits. DNS is not simply a translation language, *it is language.* It governs meaning by mandating that anything meaningful must register and appear somewhere in its system. This is the nature of protocol.

Second, as the discussion of TCP/IP shows, protocol is materially immanent. That is, protocol does not follow a model of command and control that places the commanding agent outside of that which is being commanded. It

---

28. Byfield, "DNS."

is endogenous. (This is a departure from the more hierarchical definition of protocol used by the military where control is exercised from without.)

For example, the protocological manipulation of an HTML object by an HTTP object begins first with the parsing of the HTML object:

```
<html>
<body>
Hello World!
</body>
</html>
```

The creation of a special HTTP *header* that derives from the original object is attached to the beginning of it and describes it in various ways:

```
HTTP/1.1 200 OK
Date: Sun, 28 Jan 2001 20:51:58 GMT
Server: Apache/1.3.12 (Unix)
Connection: close
Content-Type: text/html

<html>
<body>
Hello World!
</body>
</html>
```

The header contains various pieces of information about the HTML object such as the date the file was last modified (line 2), the make and model of the server offering the file (line 3), and the type of content it is (in this case, it is text-based HTML [line 5]).

The HTTP object, then, is simply the HTML object plus its new HTTP header, all wrapped up into a new form and separated by a blank line. The new header is prefixed to the original content, becoming part of its material body. But, since the HTTP header is nothing but a description of the material contents of the HTML object, the larger protocol (HTTP) is simply a way of rewriting the smaller one (HTML)—the smaller data object is encapsulated by the larger one. In doing so, the HTML object is immanently transformed—*its actual data is prefixed by another unit of data*—to function within the larger context of HTTP.

Another conclusion is that, while protocol is immanent to a particular set of data, *protocological objects never contain their own protocol.* Thus, TCP/IP houses HTTP, which houses HTML, which houses ASCII text, etc. New headers are added at each level, but in terms of content, protocols are never continuous with themselves.

At each phase shift (i.e., the shift from HTML to HTTP, or from HTTP to TCP), one is able to identify a data object from the intersection of two articulated protocols. In fact, since digital information is nothing but an undifferentiated soup of ones and zeros, data objects *are nothing* but the arbitrary drawing of boundaries that appear at the threshold of two articulated protocols.[29] In order to see HTML, one must actually view it as it intersects with HTTP. Otherwise, one looks at HTML and sees nothing but its own internal protocols: text and markup tags.

A last point, something that should be of particular interest to critical theorists, is that protocol is *against interpretation.* This is to say that protocol does little to transcode the meaning of the semantic units of value that pass in and out of its purview. It encodes and decodes these values, yes, but such transformations are simply trivial mathematics and do not affect meaning in the same way that a Hollywood film may affect the meaning of femininity, or a police officer walking the beat may affect the meaning of power in public space. Protocols do not perform any interpretation themselves; that is, they encapsulate information inside various wrappers, while remaining relatively indifferent to the content of information contained within.

The consequences of this are legion. It means that protocological analysis must focus not on the sciences of meaning (representation/interpretation/reading), but rather on the sciences of possibility (physics or logic), which I address in more detail in chapter 5 on hacking.

The limits of a protocological system and the limits of *possibility* within that system are synonymous.

---

29. This is similar to Manovich's principle of "modularity" in which every new media object is made up of independent parts, which themselves are unique independent objects. It is, in a sense, objects all the way down. See Lev Manovich, *The Language of New Media,* pp. 30–31.

To follow a protocol means that everything possible within that protocol is already at one's fingertips. Not to follow means no possibility. Thus, protocological analysis must focus on the possible and the impossible (the envelope of possibility), not a demystification of some inner meaning or "rational kernel" within technology. *Protocol is a circuit, not a sentence.*

In this chapter on physical media I have tried to describe protocol from the perspective of its real material substrate. I described the distributed network and positioned protocol as a unique governing principle within that network. I highlighted the TCP/IP suite of Internet protocols and DNS as the two most important theoretical moments for protocol—one protocol radically distributes control into autonomous agents, the other rigidly organizes control into a tree-like decentralized database.

Next, I move beyond the hard science of protocol and begin to consider it from the perspective of form. That is: How does protocol function, not as a material machine, but as an entire formal apparatus? What techniques are used by and through protocol to create various cultural objects? How can one define protocol in its most abstract sense?

These are the fundamental questions contained in chapter 2 on form, to which I now turn.

2

Form

If everything means something else, then so does technology . . .
But technology is little more than the outer emblem or symptom
by which a systemic variety of concrete situations expresses itself in
a specific variety of forms and form-problems.
—FREDRIC JAMESON, *The Geopolitical Aesthetic: Cinema and Space
in the World System*

In chapter 1, I examine the physical realm of protocol. The physical realm refers to not only the whole mass of circuits, wires, terminals, routers, and so on that constitute protocol's material layer, but also the technical software that keeps it running. I discuss distributed networks, using the U.S. interstate highway system and the global Internet as examples. I also discuss the science behind real computer protocols such as TCP/IP and DNS. It is precisely the tension between these two machinic technologies—one deterritorializing and one reterritorializing—that creates the protocological system and allows it to be so powerful.

Yet computer protocols are not just a set of technical specifications, as described in chapter 1. They are an entire *formal apparatus.* By formal apparatus I mean the totality of techniques and conventions that affect protocol at a social level, not simply a technical one. If the previous chapter was about protocol from the point of view of the systems administrator, then the current chapter is about protocol from the point of view of the webmaster. Thus, just as film theorists have analyzed the apparatus of film in terms of film form, and ideology theorists have analyzed the apparatus of ideology in terms of its formal structure, I discuss in this chapter the formal qualities of the apparatus of computer protocols.

An attention to form is helpful for it provides the critic with a yardstick with which to evaluate the political condition of certain objects of study. Thus, one may ask: Is protocol formally healthy? I attempt to answer this question at various moments throughout the book.

Writing about radio, the Marxist cultural worker Bertolt Brecht articulated a formal theory that would prove influential for theorists of communication networks of all sorts. His criticism was that radio was in the wrong shape, that it had yet to fulfill its full potential for being a two-way communications network:

Radio is one-sided when it should be two-. It is purely an apparatus for distribution, for mere sharing out. So here is a positive suggestion: change this apparatus over from distribution to communication. The radio would be the finest possible communication apparatus in public life, a vast network of pipes. That is to say, it would

---

Epigraph: Fredric Jameson, *The Geopolitical Aesthetic: Cinema and Space in the World System* (Bloomington: Indiana University Press, 1992), pp. 9, 1.

be if it knew how to receive as well as transmit, how to let the listener speak as well as hear, how to bring him into a relationship instead of isolating him.[1]

This comment has proven to be immensely clairvoyant, for not only has it foreshadowed the actual architecture of electronic networks like the Internet (which are based on both transmission and reception), it has laid the foundation for a Marxist theory of the media based on form.

But what would a Marxist theory of the media actually look like? This is the problem faced by Hans Magnus Enzensberger in his essay "Constituents of a Theory of the Media." Enzensberger is directly influenced by Brecht's desire for a two-sided radio network, where each receiver is also a transmitter. He first summarizes Brecht's position and then shows how it suggests a type of political prohibition. For why wouldn't the many-to-many radio network already exist, if it was *possible* for it to exist? Enzensberger's answer is that it is prohibited for "political reasons," and that this prohibition arises from the fundamental structure of capital: "The technical distinction between receivers and transmitters reflects the social division of labor into producers and consumers . . . It is based, in the last analysis, on the basic contradiction between the ruling class and the ruled class."[2] Thus, after lamenting that "so far, there is no Marxist theory of the media,"[3] he takes the first step to actually define media within a Marxist-materialist framework.

Realizing that power is a complex process rooted in both vertical hierarchies and horizontal networks, Enzensberger critiques Leftist positions on social manipulation (typified by his compatriots in the Frankfurt School) in which culture and consciousness emanate downward from a manipulative and alienating cultural totality. Enzensberger writes that the "unspoken basic premise of the manipulation thesis" is that "there is such a thing as pure, unmanipulated truth."[4] On the contrary, Enzensberger believes that such un-

---

1. Bertolt Brecht, "The Radio as an Apparatus of Communication," in *Video Culture,* ed. John Hanhardt (Layton, UT: Peregrine Smith Books, 1986), p. 53.

2. Hans Magnus Enzensberger, "Constituents of a Theory of the Media," in *Video Culture,* ed. John Hanhardt (Layton, UT: Peregrine Smith Books, 1986), p. 98.

3. Enzensberger, "Constituents," p. 97.

4. Enzensberger, "Constituents," p. 101.

manipulated truth is folly. And to the extent that transmission itself means being able to manipulate ("every use of the media presupposes manipulation"[5]), then *everyone* interested in an emancipated media should be a manipulator. In this sense, media are by their very nature "dirty" for they require, in the very act of critique, to engage with the dominant technologies of manipulation.

Taking his cue from Marx's sometimes sarcastic rhetoric, Enzensberger offers readers the following political warning: "Fear of handling shit is a luxury a sewerman cannot necessarily afford,"[6] meaning that those who are oppressed (the "sewerman") cannot be afraid to engage with the media (the "shit") that oppresses them. Later, Enzensberger elaborates on his criticism of the "pure" politics of the naive Left: "fear of being swallowed up by the system is a sign of weakness; it presupposes that capitalism could overcome any contradiction—a conviction which can easily be refuted historically and is theoretically untenable."[7]

With this in mind, it is clear that the media can contain emancipatory characteristics, as Enzensberger shows in the following comparison chart:

Repressive use of media	Emancipatory use of media
Centrally controlled program	Decentralized program
One transmitter, many receivers	Each receiver a potential transmitter
Immobilization of isolated individuals	Mobilization of the masses
Passive consumer behavior	Interaction of those involved, feedback
Depoliticization	A political learning process
Production by specialists	Collective production
Control by property owners or bureaucracy	Social control by self-organization[8]

The two political camps that Enzensberger describes here have historical meaning. The repressive mode is most closely associated with modern media

---

5. Enzensberger, "Constituents," p. 103.

6. Enzensberger, "Constituents," p. 101.

7. Enzensberger, "Constituents," p. 103.

8. Enzensberger, "Constituents," pp. 110–111.

(network television, film, radio), while the emancipatory mode is most closely associated with postmodern media (Internet). Indeed Enzensberger's "emancipatory" column is a near perfect foreshadowing of today's most optimistic descriptions of the Internet.

Using a philosophy of many-to-many communication derived from Brecht's essays on radio, Enzensberger equates decentralization with Marxist liberation. He lauds the new electronic media for being "oriented towards action, not contemplation; towards the present, not [bourgeois] tradition."[9] The very immateriality of the media resists commodification and reification, suggests Enzensberger: "The media produce no objects that can be hoarded and auctioned,"[10] and later: "The media also do away with the old category of works of art which can only be considered as separate objects . . . The media do not produce such objects. They create programs. Their production is in the nature of a process."[11] The discovery of processes where once there were objects—this is perhaps the most fundamental moment in a Marxist method.

Jean Baudrillard's "Requiem for the Media" was inspired by Enzensberger, but he rewrites Enzensberger's battle cry "there is no Marxist theory of the media" as simply "there is no *theory* of the media," Marxist or otherwise.[12] This suggests that Baudrillard wants to push media theory out of the realm of pure Marxism (Enzensberger's position) and into the realm of signification and communication. He says as much: "One retains the general form of Marxist analysis . . . , but admits that the classical definition of productive forces is too restricted, so one expands the analysis in terms of productive forces to the whole murky field of signification and communication."[13]

While ostensibly non-Marxist, it is worth noting here the work of Norbert Wiener and Vannevar Bush, two of the most important thinkers in the history of computers and electronic media.

---

9. Enzensberger, "Constituents," p. 105.

10. Enzensberger, "Constituents," p. 105.

11. Enzensberger, "Constituents," p. 121.

12. Jean Baudrillard, "Requiem for the Media," in *Video Culture,* ed. John Hanhardt (Layton, UT: Peregrine Smith Books, 1986), p. 124.

13. Baudrillard, "Requiem for the Media," pp. 124–125.

Wiener's theory of dynamic systems, known as cybernetics, acts as an alternative or even a precursor to network theory. The theory of cybernetics began with the simple idea of feedback. Feedback means that certain processes, having both a beginning and ending point, should be able to receive new input about their surroundings throughout their duration. The process is then able to change itself according to data received from its surroundings.

Cybernetics has become associated with man-machine hybrids (radicalized by Donna Haraway's "cyborg" theory of recent years) because, as Wiener recognized, the feedback mechanism need not be organic. Wiener considered there to be a special relationship between computers and the human brain. Thus it is a logical step to replace organic sense organs with computerized ones. In this way the cybernetic system is born. Its virtues are balance, self-regulation, circularity, and control. While Wiener's focus on systemic dynamism was certainly emulated by later network theorists, his focus on small, closed systems was not.

Bush's 1945 essay "As We May Think" is famous today for its proposed "memex" technology and other prescient ideas that prefigure much of today's networked technologies. Like Wiener, Bush considered there to be a special isomorphism between the structure of the brain and the structure of electronic technologies such as networks. He was obsessed with making technology more transparent, more like the human brain, which he believed operated by associative relationships rather than linear ones. The human mind "operates by association,"[14] he writes.

His imaginative offering was the Memex, a nonhierarchical, associative machine for inputting and outputting information. "A memex is a device in which an individual stores all his books, records, and communications, and which is mechanized so that it may be consulted with exceeding speed and flexibility. It is an enlarged intimate supplement to his memory."[15] The innovation of the Memex, however, is its architecture. It was to constitute a

---

14. Vannevar Bush, "As We May Think," in *Electronic Culture,* ed. Timothy Druckrey (New York: Aperture, 1996), p. 40.
15. Bush, "As We May Think," p. 41.

type of meshwork, a relational database of records operating on the principle of associative, rather than hierarchical, indexing.[16]

Both Wiener and Bush have therefore unwittingly contributed greatly to the tradition of Marxist media theory inaugurated by Brecht. Bush's meshworks offer a profound alternative to the centralized, hierarchical power in place under capital relations (e.g., in hierarchy within the factory). Wiener's cybernetic theory emulates several of the virtues of Enzensberger's chart of "emancipatory" media including the idea of "self-organization," the focus on process, the imbrication of input and output (reception/transmission), and the idea of "feedback" itself.

The Internet is deceptive. Critics love to exclaim that "everything has changed!" They write that advances such as new media, new technologies, new and faster methods of transferring information, democratization of technological luxuries, diversification of access to digital networks, the standardization of data formats, and the proliferation of networked relations will help usher in a new era marked by greater personal freedom, heightened interpersonal communication, ease from the burden of representation, new perspectives on the problem of the body, greater choice in consumer society, unprecedented opportunities for free expression, and, above all, speed.

Here are but a few examples, picked nearly at random, from recent literature:

Pierre Lévy—"Never before have the technological, economic, and social changes around us occurred so rapidly or been so destabilizing."[17]

---

16. Tim Berners-Lee, the inventor of the World Wide Web, was cognizant of Bush and would echo him many years later: "The vision I have for the Web is about anything being potentially connected with anything. It is a vision that provides us with new freedom, and allows us to grow faster than we ever could when we were fettered by the hierarchical classification systems into which we bound ourselves. . . . And it brings the workings of society closer to the workings of our minds." See Berners-Lee, *Weaving the Web* (New York: HarperCollins, 1999), pp. 1–2. Berners-Lee's historical acumen should not be overstated, however, as he elsewhere misspells the names of both Donald Davies and Paul Baran, the two co-inventors of packet-switching (p. 6). But, alas, "academic quality" is a "subjective notion," he reminds readers (p. 125).

17. Pierre Lévy, *Becoming Virtual: Reality in the Digital Age* (New York: Plenum, 1998), p. 16.

Machiko Kusahara—"Our culture is undergoing a truly drastic change in terms of our physical and psychological relationship with space and other bodies."[18]

Maurizio Lazzarato—"Not, perhaps, since the printing press's invention has European culture experienced so much upheaval. The very underpinnings of the notion of culture and its modes of production, socialization, and appropriation are under attack."[19]

Manuel Castells—"My starting point, and I am not alone in this assumption, is that, at the end of the twentieth century, we are living through one of these rare intervals in history . . . 'that occur with great rapidity' [Stephen J. Gould] . . . An interval characterized by the transformation of our 'material culture' by the works of a new technological paradigm organized around information technologies."[20]

But if critical theory teaches anything, it is to be wary of the obvious. These exclamations about the obvious new qualities of new media are worthy of closer examination.

The story goes that the Internet is rhizomatic. On the one hand, the Web is structured around rigid protocols that govern the transfer and representation of texts and images—so the Web isn't "an acentered, nonhierarchical, nonsignifying system" as is Deleuze and Guattari's rhizome. But on the other hand, the Web seems to mirror several of the key characteristics of the rhizome: the ability of any node to be connected to any other node, the rule of multiplicity, the ability to splinter off or graft on at any point, the rejection of a "deep structure," and so forth.

The Web is described as a free, structureless network. Yet the rhizome is clearly not the absence of structure. It is the privileging of a certain kind of structure, the horizontal network, over another structure, the tree. So to equate the Web with the rhizome, one must argue against those who describe

----

18. Machiko Kusahara, "Presence, Absence, and Knowledge in Telerobotic Art," in *The Robot in the Garden,* ed. Ken Goldberg (Cambridge: MIT Press, 2000), p. 200.

19. Maurizio Lazzarato, "New Forms of Production and Circulation of Knowledge" in *Readme!,* ed. Josephine Bosma et al. (New York: Autonomedia, 1999), p. 159.

20. Manuel Castells, *The Rise of the Network Society* (Oxford: Blackwell, 1996), p. 29.

Incontinence

The studium in this image is the massive grid, consistent and smooth. The punctum is the dead man, his incontinent golf cart having sliced a low arc across the space. Incontinence is the closet door that swings open repeatedly and without explanation in Hitchcock's *The Trouble with Harry*. The protocological matrix is incapable of restraining this type of natural expression. That is its greatest strength.

The Parallax View (1974)

the Web as a free, structureless network, and argue for a certain kind of rhizomatic protocol on the Web. This is a discussion already underway in chapter 1 and in the introduction.

The project of this book is to show that protocol is in fact both poles of this machinic movement, territorializing structure and anarchical distribution. So the question is, How exactly does protocol sew together the apparatus of digital computing into an intelligible object? Why is the Internet such an enthralling technology? How does it work so well while remaining anarchic?

Indeed, the Internet works too well. If the Internet were truly rhizomatic, it would resist identification. It would resist the deep, meaningful uses that people make of it everyday. The Net is not narrative-based, or time-based. But it still enthralls users, dragging them in, as television and cinema did before it. How? The answer is in the form.

One concept that I will borrow from film theory is *continuity*. Despite being a decentralized network composed of many different data fragments, the Internet is able to use the application layer to create a compelling, intuitive experience for the user. This is the secret of continuity.

On the Web, the browser's movement is experienced as the user's movement. The mouse movement is substituted for the user's movement. The user looks through the screen into an imaginary world, and it makes sense. The act of "surfing the Web," which, phenomenologically, should be an unnerving experience of radical dislocation—passing from a server in one city to a server in another city—could not be more pleasurable for the user. Legions of computer users live and play online with no sense of radical dislocation.

Continuity, then, is defined as the set of techniques practiced by webmasters that, taken as a totality, create this pleasurable, fluid experience for the user. As a whole they constitute a set of abstract protocological rules for the application layer.

What follows are some of the most important techniques of continuity. They will not be new to anyone already familiar with the Net. There are no RFCs to define these protocols, but they exist as dominant standards nonetheless.

Conceal the source. Many media formats have a tendency to conceal their own making. This is one reason why Marx's formal critique of the commodity form has been so useful for film theorists, because the commodity itself has a tendency to conceal its own making.

In classic Hollywood film, the apparatus is deliberately removed from the image in the same way that the process of production is removed from the commodity. Although computers are very different, a similar logic is at play. Protocol is simply a wrapper. It must conceal its own innards (or at least have a merely quantitative relationship with them via certain mathematical operations like the checksum, content length, etc.). "The job of computers and networks," writes Berners-Lee, "is to get out of the way, to not be seen. . . . The technology should be transparent, so we interact with it intuitively."[21]

Two common examples are HTML and IP addresses. HTML is the textual information that is sent to the user's browser when he or she "goes" to a Web page. This textual information is never revealed to the user; instead, it is kept hidden, interpreted as a graphical Web page. The user must voluntarily opt to view the source in order to circumvent this convention.

IP addresses are the numerical addresses given for each location on the Net. Untranslated, they come in the numerical form 12.34.56.78. Yet IP addresses are always converted into the dot-com words that make up Internet domain names such as www.amazon.com or www.etoys.com. Despite being more fundamental (i.e., all domain names are converted into numbers before a Web page is retrieved by the user's browser), the numerical addresses are subordinated to the domain names and kept hidden.

Other examples include image files that must also conceal their source. The raw code of images is never made visible to the user. Instead the code is interpreted and rendered *as an image.*

Programming languages also follow the rules of continuity: The visible code written by the programmer is made invisible at the moment the code is compiled. It is changed from a legible script into a hermetically sealed executable file (even scripting languages like Perl are interpreted before they are executed).

Eliminate dead links. On the Internet, dead links are called "404 errors." For successful continuity, 404 errors are to be avoided at all costs. This is true for any object on the Internet, be it an HTML page, an image, or a special media object (such as a Java applet). If something is pointed to, it must exist.

---

21. Berners-Lee, *Weaving the Web,* p. 159.

A dead link cuts the flow of information to the user. It is a lie. It is a corruption of the smooth space created on the Web.

Eliminate no links. As important to continuity as the elimination of dead links is the elimination of no links. There can be no dead ends on the Internet. Each page *must go somewhere else,* even if that somewhere else is "back." Each page must be rich with offerings, it must situate itself within an abundant network of flows. It must open doors to the user, offering free channels of infinite movement.

Green means go.[22] When creating links on a Web page, representation should convey meaning. That is, the link itself should convey to the user where he or she will go upon following the link. "Green means go" is just that: A green traffic light intuitively means "to go." Capitalize on the user's preexisting knowledge of iconography to designate a link. "Green means go" means that one must never use the phrase "click here to visit . . ." Instead, for optimal continuity, one should inject the meaning of the link directly into its form. As Steven Johnson has noted, "Good interfaces do away with text, the way good psychotherapists get rid of repressed memories and emotional blocks."[23]

True identity. Worse than a dead link is a deceptive link. A link's name and its address must correspond. Wherever the user goes, there he or she must be. If the link's name and its address do not correspond, then a more insidious type of discontinuity is in effect than even the 404 error could provoke. For while the flows have not been interrupted, they have been infected with disinformation. One flow has been substituted for another flow.

Remove barriers. Each click that a user is forced to make on the Web is an unnecessary barrier to that user and will hinder his or her movement. All unnecessary barriers (splash pages, tables of contents, introductory pages) between the user and the content must be removed. All content must be placed as close to the front door as possible. Often this means that the tree-like hierarchy of computer operating systems must be subverted, in favor of a more flat interface.

---

22. This expression is taken from Jeffrey Veen, *Hotwired Style* (San Francisco: Wired Books, 1997), pp. 54–55.

23. Steven Johnson, *Interface Culture* (New York: Basic Books, 1997), p. 150.

Continuity between media types. This principle is crucial, and I discuss it in greater detail later. In essence, all differentiation between different types of media—text, images, animations—must be eliminated.[24] The digital nature of information makes this possible.

Smooth transitions between text and image are important. For example, on the Internet the *page* itself is important, not its component parts. The user must not be able to tell where one image begins and another ends, or where text ends and an image begins.

Prohibition against low resolution. All graphics must aspire to wholeness. The technique called anti-aliasing, whereby jagged shapes are made fuzzy in order to increase the image's overall resolution, is very important in this process. Pixels cannot be visible. Fonts must be smooth. Full color palettes are better than limited ones. High-resolution, representational images are better than cheap, simple ones. Low resolution shatters the illusion of continuity because it means that the source, the code, is not being properly concealed.

Consider ASCII, for example. The American Standard character set known as ASCII is synonymous with low resolution. The use of ASCII characters in various graphical techniques has become synonymous with low resolution (a technique I mention in chapter 7). Graphical browsers, which can produce vivid Web pages, are privileged over text-based browsers, which are limited to the ASCII character set. HTTP is privileged over other, nongraphical protocols such as telnet.

Highest speed possible. In both processor speed and network bandwidth, faster is always better. Speed means continuity; lack of speed means discontinuity. Speed helps perpetuate the illusion that personal movement on the Net is unmediated, that the computer is a natural extension of the user's own body.

Prohibition on crashes. Computers must never crash. A downed server is the ultimate disruption of Net continuity. Not only does the crash disrupt the movement of the user, it is offensive, attacking the user itself with threats of data loss and software corruption. During a crash, the computer changes from being passive to being active.

---

24. Pierre Lévy writes: "All representation is now subject to sampling, mixing, reuse, etc. According to the pragmatics of creation and emerging communication, nomadic distributions of information will fluctuate around an immense *deterritorialized semiotic plane.*" See Lévy, *L'intelligence collective: Pour une anthropologie du cyberspace* (Paris: Éditions la Découverte, 1994), p. 122.

Prohibition on dead media. All technology is new technology. Everything else must be eliminated. Dead media are those media that have fallen out of use. They include everything from dead hardware (Commodore 64, Apple II) to dead computer languages (Fortran, Cobol) to dead media formats (Sony Betamax, 8-track cassette). Dead media means discontinuity. Dead media do not function within contemporary network flows; therefore, they have no place.

Eliminate mediation. The goal of continuity is to make the Internet as intuitive as possible, to make the network a natural-feeling extension of the user's own body. Thus, any mediation between the user and the network must be eliminated. Interfaces must be as transparent as possible. The user must be able to move through the network with unfettered ease.

All traces of the medium should be hidden, hence the evolution from the less intuitive "QWERTY" keyboard to technologies such as the touch screen (e.g., Palm and other PDAs) and voice recognition software.

Feedback loops. As the discussion of Brecht and Enzensberger shows, the history of media has been the history of the prohibition of many-to-many communication. Many-to-many communication is a structure of communication where each receiver of information is also potentially a sender of information. The Internet with its email and open webpage architecture, is one of the few counterexamples to this tendency in mass communication.[25] Feedback loops are necessary to help produce the active subjectivity of the user.

Thus, on the Internet there exists a proliferation of feedback loops such as email responses, input forms, online chat rooms, and bulletin boards. They produce a many-to-many communication effect in the realm of signification that is central to continuity.

---

25. One of the most insidious new methods for limiting many-to-many communication on the Internet is the asynchronous bandwidth systems available in today's popular Digital Subscriber Lines (DSLs) and cable modems. Traditionally, communication between two computers was always synchronous; that is, both outgoing and incoming information could travel at the same speed. However, as is common with DSL and cable modem systems, downstream bandwidth (the size of the pipe entering your house) can be as much as ten times the size of the upstream bandwidth (the size of the pipe leaving your house). Thus, it becomes easier to receive information than it does to send it, creating a scenario much more like television than the many-to-many structure promised by the Internet.

Anonymous but descriptive. The conflict between the total and the specific is palpable on the Internet. Each movement on the Net is recorded in myriad different locations (log files, server statistics, email boxes); however, the real identity of those movements is irrelevant. Demographics and user statistics are more important than real names and real identities. On the Internet there is no reason to know the name of a particular user, only to know what that user likes, where they shop, where they live, and so on. The clustering of descriptive information around a specific user becomes sufficient to explain the identity of that user.

Foucault introduced the concept of "biopower" to help explain this phenomenon. His formulation was consistent with the functioning of protocol, for biopower is the power to interpret material objects as information, to affect objects at the statistical or informational level, not at the level of individual content.

Just as the "180-degree rule" is to the conventions of Hollywood film, several of the techniques should be observed in order to conform to the formal protocols of the Net.

Of course, the most interesting forms of cultural production appear when some of these principles are inverted, a practice addressed in part III.

As this quick overview of Web continuity suggests, the Net is not simply a new, anarchical media format, ushering in the virtues of diversity and multiplicity, but is, in fact, a highly sophisticated system of rules and regulations (protocol).

For many years now theorists have preferred to speak of value economies—be they semiotic or economic—in terms of genetic units of value and the general equivalents that regulate their production, exchange, and representation. Yet the Net has a different, more horizontal system of command and control. Tempting as it may be to follow the lead of film critics like Christian Metz and André Bazin and claim that, like cinema before it, the whole of digital media is essentially a language, or to follow the lead of Tel Quel Marxist Jean-Joseph Goux (or even the early economics-crazed Baudrillard) and claim that digital media are essentially value economies regulated by the digital standard of ones and zeros, it is clear that digital media require a different kind of semiotics, or perhaps something else altogether.

The Net does not rely on the text as its primary metaphor; it is not based on value exchange; it is not time-based like film or video; it is not narrative

A packet radio test van at SRI
International in 1979 with the
Army 18th Airborne Corps. Packet
radio, ARPAnet, and packet
satellite nets were the first three
networks making up the Internet.

Vint Cerf in Johannesburg, South
Africa, in 1974 demonstrating the
ARPAnet. This was the first such
demonstration of ARPAnet in
South Africa. It was done in
conjunction with a conference
sponsored by the International
Federation of Information
Processing Societies (IFIPS).

in a conventional sense; its terms are not produced in a differential relationship to some sort of universal equivalent. Digital technology necessitates a different set of object relations. The previous points are an attempt to describe these relations.

I turn now from my introduction to the creation of continuity in Net form to a more abstract consideration of formal protocol. As described in chapter 1, the physical part of the Net apparatus is its hardware. There are many different types of hardware: controllers (keyboards, joysticks), virtualization apparatuses (computer monitors, displays, virtual reality hardware), the interface itself (i.e., the confluence of the controller and the virtualization apparatus), the motherboard, and physical networks both intra (a computer's own guts) and inter (an Ethernet LAN, the Internet). However, the niceties of hardware design are less important than the immaterial software existing within it. For, as Alan Turing demonstrated at the dawn of the computer age, the important characteristic of a computer is that it can mimic *any machine,* any piece of hardware, provided that the functionality of that hardware can be broken down into logical processes. Thus, the key to protocol's formal relations is in the realm of the immaterial software.

Record

The first term in Net form is the record. The record has its roots in the ability of physical objects to store information. A record is any type of nonrandom information, not simply something that records language or data. Thus, the act of sharpening a raw stone into a tool embodies the stone with the "information" of its new shape. Arranging randomly scattered leaves into a straight line gives the leaves "information."

As Vilém Flusser notes, different physical objects have different propensities for storing information:

Air has the advantage of being readily accessible; moreover, we have organs[26] which seem to have been made to transform airwaves into signs (to make "phonemes" out of them). . . . Hard objects (stones and bones) have the advantage of storing information recorded in them for a relatively long time. . . . Approximately three thou-

---

26. Namely, our ears and vocal chords.

sand five hundred years ago (in other words, only a short time ago), an important step was taken; the alphabet was invented. It is a system which recodes the phonemes of spoken languages into visual signs, allowing them to be engraved into hard objects.[27]

Certain records can experience a conjunction of utility and information. Thus, a knife not only contains the information of cutting in its form, but is also used to cut. A photograph of a knife, on the other hand, contains the information of cutting, but cannot be used to cut.

With the alphabet comes a perfect synergy of form and information. Not only does the inscription of language have a meaning, it records that meaning in the very act of its own inscription. Then, as Kittler has observed, at a certain historical moment the inscription of language was bifurcated into two semiotic entities, the material object of storage and the meaning to be stored. Looking at the "moment" of 1900 (the moment of the phonograph and the typewriter), he writes that "the ability to record sense data technologically," using such instruments as the phonograph and the typwriter, "shifted the entire discourse network circa 1900. For the first time in history, writing ceased to be synonymous with the serial storage of data. The technological recording of the real entered into competition with the symbolic registration of the Symbolic."[28]

This shift was also observed by Ferdinand de Saussure who, in his lectures that would make up the *General Course in Linguistics,* labeled the material object a "signifier" and the meaning contained in it a "signified." The record is, in the most abstract sense, any nonchaotic *something.*

Object

A record is one particular form-of-appearance of an *object.* The object is the digital economy's basic unit. It is any unit of content. It is not simply a digitization of the Marxist commodity, or a digitization of the semiotic sign. The object is not a unit of value. "A new media object," writes Lev

---

27. Vilém Flusser, "Memories," in *Ars Electronica,* ed. Timothy Druckrey (Cambridge: MIT Press, 1999), p. 203.

28. Friedrich Kittler, *Discourse Networks, 1800/1900,* trans. Michael Metteer (Stanford: Stanford University Press, 1990), pp. 229–230.

Manovich, "may be a digital still, digitally composed film, virtual 3-D environment, computer game, self-contained hypermedia DVD, hypermedia Web site, or the Web as a whole."[29] But I would take that even further and say that the digital object is any positive content-unit or content-description: text, image, MIDI data, VRML world, texture, movement, behavior, transformation. Digital objects are pure positivities. They are the heterogenous elements that exist in what Deleuze and Guattari have called "machinic" processes.

These objects are always derived from a preexisting copy (loaded) using various kinds of mediative machinery (disk drives, network transfers). They are displayed using various kinds of virtuation apparatuses (computer monitors, displays, virtual reality hardware). They are cached. And finally, objects always disappear.

Objects exist only upon use. They are assembled from scratch each time and are simply a coalescing (of their own objectness). Unlike the Marxist commodity and the semiotic sign, the object is radically independent from context. Objects are inheritable, extendible, procreative. They are always already children. Objects are not archived, they are autosaved. Objects are not read, they are scanned, parsed, concatenated, and split.

Different objects are understood as such due to their irreconcilability, their separation within a machinic process. Thus one might arrange a table of the digital objects existing on the average computer desktop:

filename	↔	extension
identity	↔	datatype
data	↔	format
page	↔	link

Protocol

As shown here, a protocol is a set of rules that defines a technical standard. But from a formal perspective, protocol is a type of object. It is a very special kind of object. Protocol is a universal description language for objects.

Protocol is a language that regulates flow, directs netspace, codes relationships, and connects life-forms. Protocol does not produce or causally effect objects, but

---

29. Lev Manovich, *The Language of New Media* (Cambridge: MIT, 2001), p. 14.

rather is a structuring agent that appears as the result of a set of object dispositions. Protocol is the reason that the Internet works and performs work. In the same way that computer fonts regulate the representation of text, protocol may be defined as a set of instructions for the compilation and interaction of objects. Protocol is always a second-order process; it governs the architecture of the architecture of objects. Protocol is how control exists after distribution achieves hegemony as a formal diagram. It is etiquette for autonomous agents. It is the *chivalry* of the object.

The Internet is a delicate dance between control and freedom. As Gerfried Stocker and Christine Schöpf have noted:

A conception of media oriented upon transmission and dissemination (that is, centralized, unidirectional distribution) has become passé in actual artistic practice. This conceptual schema—one rooted in the industrial epoch and in which the overcoming of geographical distance, the transfer of messages, and thus speed are inherent central parameters—is now countered by the concept of omnidirectional and participatory spheres of communication of which the Internet is the prototypical example.[30]

In other words, at the same time that it is distributed and omnidirectional, the digital network is hegemonic by nature; that is, digital networks are structured on a negotiated dominance of certain flows over other flows. Protocol is this hegemony. Protocol is the synthesis of this struggle.

Browser
One of the defining features of intelligent networks (capitalism, Hollywood, language) is an ability to produce an apparatus to hide the apparatus. For capitalism, this logic is found in the commodity form; for Hollywood, it is continuity editing. In digital space this "hiding machine," this making-no-difference machine, is epitomized in the Internet browser.

Despite recent talk about the revolutionary potential of artist-produced browsers (Web Stalker[31] is the first and most famous example), I consider all

---

30. Gerfried Stocker and Christine Schöpf, "Preface," in *Ars Electronica,* ed. Timothy Druckrey (Cambridge: MIT Press, 1999), p. 14.

31. See http://www.backspace.org/iod.

browsers to be functionally similar and subdivide them into the following categories: dominant (Mosaic, Netscape, Explorer, Neoplanet, Opera, etc.), primitive (Lynx), special media (VRML browsers, Applet viewers, audio/video players, etc.), and tactical (Web Stalker, Netomat, etc.). While the Net has existed already for decades, it is only recently that more sophisticated browsers have emerged out of earlier, primitive software. Paralleling the emerging dominance of windows-style operating systems (MacOS, Microsoft Windows) over text-based operating systems (UNIX, DOS), the browser slowly evolved from its primitive text-based form into the graphical browsers of today. Graphical browsers are highly complex protocological objects.

As I said in chapter 1, the goal of protocol is totality, to accept everything. This principle is also exhibited in the browser. Its goal is to display all media formats. The browser is an interpreting apparatus, one that interprets HTML (in addition to many other protocols and media formats) to include, exclude, and organize content. It is a valve, an assembler, a machinic process.

In the browser window, data objects (images, text, etc.) are pulled together from disparate sources and arranged all at once each time the user makes a request. The browser is fundamentally a kind of filter. It is a machine that uses a set of instructions (HTML) to include, exclude, and organize content. Its virtue is not diversity but *university*.

HTML

As the Net's universal graphic design protocol since its introduction in 1990, HTML designates the arrangement of objects in a browser. HTML is a way of marking up text files with basic layout instructions—put this sentence in boldface, add an image here, indent this paragraph—so that it is legible to a wide variety of computers and operating systems on the Web. Every Web page on the World Wide Web uses HTML.

The most important quality of HTML is that it is text only. It contains no tables, no font faces, no pictures. Yet it contains the *instructions* for tables, fonts, and pictures. For example, the following sentence has been rendered in both bold and italicized typefaces:

This is **bold type** and this is *italicized type*.

Yet if it were converted to the HTML protocol, it would look like this:

```
This is <b>bold type</b> and this is <i>italicized
type</i>.
```

In HTML, "<b>" stands for bold and "<i>" stands for italics. But note that
in HTML the words "bold type" and "italicized type" are not actually rendered in bold or italics; they are simply wrapped in protocological "tags"
that designate bold or italics. The final design layout is never actually included in the HTML file; it is merely *described* through a series of tags.

While HTML does require more typing, it actually simplifies graphic
layout by breaking it into standard textual instructions. Why? Two reasons:
(1) on the Internet, plain text is the quickest type of data object to download,
and (2) a shared standard is necessary for data interchange between many different types of computers.

As the HTML specifications note, "to publish information for global distribution, one needs a universally understood language, a kind of publishing
mother tongue that all computers may potentially understand."[32] HTML is
therefore nothing more than a protocol for graphic design. As a protocol, it
facilitates similar interfacing of dissimilar objects.

Fonts

A font is not analogous to a signifier. Rather it renders the signifier itself internally complex. It is a subelement of the signifier. A computer font cannot
be thought of, therefore, as a genetic element of the sign. In text, for example,
a font must be thought of independently from content, written markings,
and so forth. Fonts are protocological. They regulate representation. Font
faces appear at the intersection. They are the veneer of representation. The
font is always the first thing read and the last thing written. Fonts have no
body, only a formation. They buffer the act of reading. They protect the
reader from the shock of virtual transfer. They are a formal protocol.

Computer fonts do the same work in the digito-semiotic world that
HTML does in the virtual world. Both are a set of instructions for the compilation of contents. Fonts compile and represent digitized texts, while HTML
compiles and displays hypertextual elements. Like HTML, a computer font

---

32. From the HTML specification sheet, available online at http://www.w3.org.

displays textual information "all at once," and virtually. On load, a derivative of each element is placed. On unload, that copy is discarded. However, computer fonts are not representation per se. They are governing principles for representation. They are at once totally crucial to the transfer of textual information and yet they are completely disposable, contingent, and atemporal. They are a readable example of protocol.

In this chapter I have attempted to analyze the formal protocols that inhabit what Gene Youngblood calls the "intermedia network," that is, the entire apparatus of signification that surrounds distributed computer networks. Like Marx's analysis of the commodity, or Bazin's analysis of film form, Net form must be decoded to reveal its inner complexities.

I have addressed several of these complexities—techniques like "conceal the source" or "prohibition on dead media"—that contribute to an overall sense of continuity achieved, however flimsy it may appear to some, in the space of the network. I have also described the genealogy from record to object and finally to protocol that created the conditions of existence for several highly protocological machines such as HTML or the browser itself.

Thus, having discussed the physical science of protocol in chapter 1, and the formal qualities of protocol in this chapter, I move now to the subject of power, or protocol as a type of distributed management system for both human and nonhuman agents.

3

Power

Technology is social before it is technical.
—GILLES DELEUZE, *Foucault*

Thus far, I have considered protocol in both its physical and formal capacities. In the previous chapter I showed that protocol consists of an entire formal apparatus, one that encapsulates information inside various cultural wrappers exemplified by the Internet browser, the markup language known as HTML, among others. These wrappers tend to be ignorant of their contents. "That's the trouble with bits," writes computer guru Charles Petzold. "They're just zeros and ones and don't tell you anything *about* themselves."[1] (Even smart markup languages like XML require the intervention of human brains to establish semantic contexts.) These wrappers control their contents through inflection, connectivity, contextualization, and other protocological techniques.

It is now time to consider protocol in its political sense, as a pseudo-ideological force that has influence over real human lives. Recall, then, Deleuze's "control societies" described first in the introduction. Deleuze defines control societies as being primarily "digital." They operate through "ultrarapid forms of apparently free-floating control."[2] In a similar vein, Michel Foucault has argued that the further the progression into the postmodern (or digital) age, the more politics ceases to interest itself in the soul or the body. Instead, politics desires "life itself." Foucault calls this type of politics "bio-politics."

I argue in this chapter that protocol has a close connection to both Deleuze's concept of "control" and Foucault's concept of biopolitics. I show here that protocol is an affective, aesthetic force that has control over "life itself." This is the key to thinking of protocol as power.

Within the protocological system, life engenders and molds itself into a type of social sculpture (Beuys). Protocol never operates *in general,* but instead exists through specific material management styles that are made visible

---

Epigraphs: Gilles Deleuze, *Foucault,* trans. Seán Hand (Minneapolis: University of Minnesota Press, 1986), p. 40. Georg von Lukács, "Thoughts Toward an Aesthetic of the Cinema," trans. Janelle Blankenship, *Polygraph* 13 (2001), p. 16. Frederick Cohen, *A Short Course on Computer Viruses* (New York: John Wiley & Sons, 1994), p. 20.

1. Charles Petzold, *Code: The Hidden Language of Computer Hardware and Software* (Redmond, WA: Microsoft Press, 2000), p. 154.

2. Gilles Deleuze, *Negotiations,* trans. Martin Joughin (Minneapolis: University of Minnesota Press, 1990), p. 178.

within the institution of the digital network. Nonhuman life forms, referred to by artists and scientists alike as "artificial life," exist already today in such digital networks. It is the goal of this chapter to explore the political landscape on which such artificial life forms exist.

Before examining protocol as power, let me restate the definition of protocol as it has thus far been considered.

- Protocol is a system of distributed management.
- Protocol facilitates peer-to-peer relationships between autonomous entities.
- Protocol is anti-hierarchy and anti-authority.
- Protocol engenders localized decision making, not centralized.
- Protocol is robust, flexible, and universal.
- Protocol can accommodate massive contingency.
- Protocol is the outcome (not the antecedent) of distributed behavior.

That said, I attempt to argue one additional thing in this chapter: that *life,* hitherto considered an effuse, immaterial essence, has *become matter,* due to its increased imbrication with protocol forces (via DNA, biopower, and so on discussed later).

This central observation has a few corollaries: first, it is clear that for many years now *matter has become life,* this coinciding with the emergence of autonomous life forms both nonhuman and hybrid such as robots, cyborgs, and artificial life systems; second, protocol is a system of management that only exists in a space populated by a multitude of independent, vital agents; third, because protocol is agent-specific, it must always be connected to the particular material milieu inhabited by those agents—their spaces and their own material bodies. I touch upon these corollaries in what follows.

In his short essay "Society Must Be Defended," Foucault writes that "[i]n order to conduct a concrete analysis of power relations, one would have to abandon the juridical notion of sovereignty."[3] By this he means that it is impossible to explain forces of social control (from a materialist, or "concrete,"

---

3. Michel Foucault, *Ethics: Subjectivity and Truth—The Essential Works of Michel Foucault, Volume 1* (New York: The New Press, 1997), p. 59.

perspective) if one assumes that individual actors are endowed with primary rights or powers that they then *express* as political actors. Instead of studying the genetic terms of the political relationship, suggests Foucault, one must study the relationship itself.

Vilém Flusser corroborates this position when he writes:

We will have to reformulate many (maybe all) of the categories we have used until now. An example: we will have to replace the category of "subject-object" with the category of "intersubjectivity," which will invalidate the distinction between science and art: science will emerge as an intersubjective fiction, art as an intersubjective discipline in the search for knowledge; thus science will become a form of art and art a variant of the sciences.[4]

This tendency in Foucault—to diminish the importance of individual expressive actors—may be summarized as Foucault's desire to be anti-anthropological. In his historical genealogies, Foucault aspires to write a theory of history that is not based on living human beings. For example, in *Archaeology of Knowledge* Foucault expresses his desire to "define a method of historical analysis freed from the anthropological theme . . . a method of analysis purged of all anthropomorphism."[5] He claims that he wants to uncover the principles of an "autochthonic transformation"—that is, a transformation in the realm of words and things that is immanent, particular, spontaneous, *and anonymous*.

The anti-anthropomorphic tendency is also seen in his definition of discourse: "Discourse is not the majestically unfolding manifestation of a thinking, knowing, speaking subject, but, on the contrary, a *totality*, in which the dispersion of the subject and his discontinuity with himself may be determined."[6] In fact, Foucault writes that his interest lies in the "anonymous and general subject of history,"[7] not the *social* subject of history. Foucault's goal

---

4. Vilém Flusser, "Memories," in *Ars Electronica*, ed. Timothy Druckrey (Cambridge: MIT Press, 1999), p. 206.

5. Michel Foucault, *Archaeology of Knowledge* (New York: Pantheon, 1999), p. 16.

6. Foucault, *Archaeology of Knowledge*, p. 55.

7. Foucault, *Archaeology of Knowledge*, p. 16.

is, in a sense, to seize upon a history filled with mostly dead objects and animate them: "In short," he writes, "[what] we wish to do is to dispense with 'things'. To 'depresentify' them. To conjure up their rich, heavy, immediate plenitude."[8] Foucault concedes that "what is discovered by the analysis of [discursive] formations is not the bubbling source of life itself, life in an as yet uncaptured state," but rather a form of life that is the *agent* of its vitality, that creates its own vitality through the processes of living.

Indeed Foucault defines "life" in a fashion very similar to power itself. So similar, in fact, that in the late Foucault, the two terms merge into one: biopower.

In *The History of Sexuality* Foucault contrasts the older power of the sovereign over life (one characterized by the metaphysical concern of either the absence or presence of life), to a new mode in which life is either created or destroyed: "One might say that the ancient right to *take* life or *let* live was replaced by a power to *foster* life or *disallow* it to the point of death."[9] He continues: "The old power of death that symbolized sovereign power was now carefully supplanted by the *administration of bodies* and the *calculated management of life.*"[10] Foucault argues that the deployment of sexuality in the modern age is a perfect example of this type of "calculated management of life," for human lives themselves are engendered with a real, material sexuality during this period. He writes that *life enters history* at this moment—life was not a viable category before.[11] "Power would no longer be dealing simply with legal subject over whom the ultimate dominion was death, but with *living beings,* and the mastery it would be able to exercise over them would have to be applied at the level of life itself."[12] Foucault lists the achievements of biopower: "[it gives] rise to the infinitesimal surveillances, permanent controls, extremely meticulous orderings of space, indeterminate

---

8. Foucault, *Archaeology of Knowledge,* p. 47.

9. Michel Foucault, *The History of Sexuality: An Introduction, Vol. 1* (New York: Vintage, 1999), p. 138.

10. Foucault, *The History of Sexuality,* pp. 139–140, emphasis mine.

11. Foucault, *The History of Sexuality,* pp. 141–142.

12. Foucault, *The History of Sexuality,* pp. 142–143.

medical or psychological examinations, to an entire micro-power concerned with the body."[13]

A second term introduced by Foucault, biopolitics, has a slightly different meaning. Biopolitics is "the endeavor, begun in the eighteenth century, to rationalize the problems presented to governmental practice by the phenomena characteristic of a group of living human beings constituted as a population: health, sanitation, birthrate, longevity, race."[14] Elsewhere he elaborates that biopolitics "tends to treat the 'population' as a mass of living and coexisting beings who present particular biological and pathological traits and who thus come under specific knowledge and technologies."[15] Biopolitics, then, connects to a certain statistical knowledge about populations.

Foucault notes that it is dependent on the basic tenets of liberalism—namely, that people and their societies possess various statistical properties that can be measured. Biopolitics is a species-level knowledge. It is the type of knowledge that, say, UNICEF employs when it claims that the infant mortality rate in the United States is 7 per every 1,000 births.[16] This is the moment of biopolitics.[17]

---

13. Foucault, *The History of Sexuality,* pp. 145–146. In a certain sense, the most important word in this section of *The History of Sexuality* is the word "eugenics" (p.148). In Foucault's opinion, "the medicine of perversions and the programs of eugenics were the two great innovations in the technology of sex of the second half of the nineteeth century" (p. 118). So on the one hand Foucault offers perversion, and on the other hand an interest in breeding, hereditary, lineage, blood—all central concerns of the popular eugenics movements of the nineteenth and early twentieth centuries. The focus on eugenics in this section is crucial because it sets the stakes for biopower, it says that biopower really is about the control over life—be it fascist or pharmaceutical. Deleuze knew this too, writing about "pharmaceutical products, nuclear technology, and genetic engineering." See Deleuze, *Negotiations,* p. 178.

14. Foucault, *Ethics,* p. 73.

15. Foucault, *Ethics,* p. 71.

16. See http://www.unicef.org/statis. This statistic was posted on December 26, 2000, and it derives from data from UNICEF, United Nations Population Division, United Nations Statistical Division, World Bank, and U.S. Bureau of the Census—all biopolitical organizations.

17. The birth of biopolitics in the United States can probably be traced to the 1890 census. For that year's count, statistician Herman Hollerith developed a punch-card system that was

Ironically, while Deleuze is arguably Foucault's greatest champion, it is he who ultimately renders the anti-anthropological Foucault impotent, and, in a way, begins the process of *dating* Foucault. This comes late, in the "Postscript on Control Societies" (which first appeared in 1990). As discussed in the Introduction, Deleuze defines two historical periods in his essay: (1) the "disciplinary societies" of the modern age characterized by the rule of the sovereign, the "vast spaces of enclosure," the social castings and bodily molds that Foucault described so well, and (2) what Deleuze terms the "societies of control" that inhabit the late twentieth century—these are based on protocols, logics of "modulation," and the "ultrarapid forms of free-floating control"—words that to Deleuze in 1990 must have been bordering on pure science fiction. The disciplinary societies are characterized by the signature and the document, while the societies of control are characterized by the password and the computer.

Let me highlight one particular quality of the societies of control: their ability to engender nonorganic life—and this is Deleuze's intervention into the anti-anthropological concept of "subjectless patterning" I discuss earlier.

It turns on Deleuze's concept of *"dividuals."* He writes, "in control societies . . . the key thing is no longer a signature or number but a code: codes are *passwords,* whereas disciplinary societies are ruled (when it comes to integration by resistance) by *precepts.* The digital language of control is made of codes indicating where access to some information should be allowed or denied. We're no longer dealing with a duality of mass and individual" from the modern era. Instead, "individuals become *'dividuals,'* and masses become samples, data, markets, or *'banks.'*"[18]

---

processed by machine: "Hollerith's plan involved manila punch cards 6 5/8 × 3 1/4 inches in size. . . . The holes in these cards were organized into 24 columns of 12 positions each, for a total of 288 positions. These positions represented certain characteristics of a person being tallied in the census." See Petzold, *Code,* p. 241.

18. Deleuze, *Negotiations,* p. 180. In film one sees the pure, subjective individual inside the fetishistic camera play of *Picnic at Hanging Rock* (1975)—where Anne Louise Lambert's close-ups are pure pornography—or the social experiments of *Pygmalion* (1938) or Peter Brook's *Lord of the Flies* (1963). In these films the narrative unfolds in a strata dominated by affect, physical discipline, personal behavior, self-transformation, societal position, and training (or the lack

(While this essay is both exciting and disappointingly underdeveloped, what I want to point out is how it positions Foucault. For it is not simply Foucault's histories, but Foucault himself that is left behind by the societies of control. Foucault is the rhetorical stand-in for the modern disciplinary societies, while Deleuze claims to speak about the future. That is to say, in this essay Deleuze seals Foucault's fate *as theorizer of the modern,* and with it the potential for Foucault to adequately represent anti-anthropological, or protocological, thought.)

To sum up, biopolitics and biopower are Foucault's terms for protocol as it relates to life forms. They are Foucault's terms for the statistical coding, the *making-statistical,* of large living masses, such that any singular life-form within that mass may be compared in its organic nature to the totality. This is exactly how protocol functions, as a management style for distributed masses of autonomous agents.

Second Nature

The protocological management of life itself has an extensive prehistory. Foucault and Deleuze show how protocol exists today, yet life itself underwent many transformations during the modern era before it reached its current condition. As one of the greatest analysts of modern life, Karl Marx was intimately familiar with many of the sociopolitical transformations that prefigure the emergence of protocol. For example, the Marxist theory of reification is based, most simply, on a fear of life becoming matter, while, on the other hand, the theory of commodity fetishism shows vividly how matter can become alive. While these two transformations appear to move in opposing directions, they in fact foreshadow the existence of protocol.

Marx learned about materialism from the other Young Hegelians (particularly Feuerbach). He learned that the material or "natural" event that Hegel separated from spirit (i.e., from thought or idealism) must be thoroughly contemplated in its own right. "We must bore through [spiritual thinking]

---

thereof). Contrast this with "dividualized" films like Fuller's *The Naked Kiss* (1964), which reeks of a tabloid thinness and whose superficiality extends it decades into the future to a time devoid of wound-up interiority, and not at all afraid to traipse out in full view the depravities of real life. That is to say, today.

to the very foundations of truth,"[19] he wrote. These foundations are the same foundations from which protocol springs. Yet matter does not simply mean bodies and dirt and steel in Marx, for all of social life is intimately connected to the material realities of a given sociopolitical space. Indeed, as many of the "culturalist" Marxists (from Louis Althusser to Stuart Hall) have shown, the immaterial realm of cultural is as worthy of analysis as the material bases of society. In Marx's *Capital,* this "second" immaterial layer, this patina that is at once the thing itself *and* its artificial semblance, is introduced via what I term the concept of "second nature."

Other thinkers have also identified this current. John Rajchman, talking about Foucault, writes that "[i]t is not the nature of the subject that is at issue, but its '*second nature*'; not what is given, but what allows the subject the possibility of giving itself."[20]

Hannes Leopoldseder writes that "second nature" is entwined with the concept of artificiality in general:

In the year 1991, the computer lost its innocence. On January 17, 1991, at 1.00 a.m. Central European time, to be exact, when the first laser-controlled bomb met its target, the Gulf War had started . . . The Gulf War, though, did not actually create the megatrend towards artificiality, but shifted it from a symposium topic into reality, even more clearly pointing out the begin [sic] of a new myth: the myth of the artificial, *of a second Nature.*[21]

For my purposes, "second nature" refers to the way in which material objects in the modern era have a tendency to become aesthetic objects. Through being aesthetized, they also tend to become autonomous, living entities in some basic sense. This tendency is a necessary precondition for protocol, and I would like to look closely at Marx's *Capital* to illustrate how this happens.

---

19. See Karl Löwith, *From Hegel to Nietzsche* (New York: Columbia University Press, 1964), p. 73.

20. John Rajchman, *Philosophical Events: Essays of the '80s* (New York: Columbia University Press, 1991), p. 62, emphasis mine.

21. Hannes Leopoldseder, "Forward," in *Ars Electronica,* ed. Timothy Druckrey (Cambridge: MIT Press, 1999), p. 8, emphasis mine.

Indeed, Marx can help me derive my theory of protocol by showing the origins of "aesthetic" materialism.

Consider the following thoughts taken from filmmaker Sergei Eisenstein's October 13, 1927, notebook entry, as he contemplates his ambitious (but unrealized) dream of turning Marx's *Capital* into a feature film:

Assuming that in any film work, certain salient phrases are given importance, the form of a **discursive** film provides, apart from its unique renewal of strategies, their rationalization which takes these strategies into account. Here's a point of contact already with completely new film perspectives and with the glimmers of possibilities to be realized in CAPITAL, a new work on a libretto by Karl Marx. A film treatise.[22]

Discursive film . . . a *libretto* by Marx . . . a film *treatise*! How could Eisenstein transmute philosophical concepts into moving images? Does *Capital* have an aesthetic essence accessible to the medium of film? How would it be possible to make such a film, and who would watch it? From a contemporary perspective, translating a decades-old work of economics into a feature film seems impossible, even absurd.

Yet the more one considers the relationship between *Capital* and aesthetic production, the more a film version of Marx's classic text *seems like second nature.* As Annette Michelson has indicated, the "upside-downness" of life under capitalism is likened by Marx himself to both the inverted image projected on the retina and the inverted image within a *camera obscura.*[23] The tendency to invert life is shared by both the visual apparatus and the capitalist one. The cinema, then, following a path from text through the mediating space of discourse to film, seems to be an ideal form through which to represent the concept of *Capital.*

If *Capital* has a single concept, then it is the concept of the dialectical method itself. Upon close scrutiny of Eisenstein, that much is clear. The film version of "CAPITAL develops as visual instruction in the dialectical method,"

---

22. Sergei Eisenstein, "Notes for a Film of *Capital,*" *October* 2, Summer 1976, p. 4.

23. Annette Michelson, "Reading Eisenstein Reading *Capital*" (part one), *October* 2, Summer 1976, p. 27.

he writes in an April 6, 1928, notebook entry.[24] And a few days earlier: "The content of CAPITAL (its aim) is now formulated: **to teach the worker to think dialectically.** / To show the **method** of dialectics."[25] Eisenstein considered his work to be closer in form to an essay than a conventional film. And, as his interest in montage matured, the dialectical method remained his central problematic. Michelson also points out that "Eisenstein was at pains to ground . . . montage in the dynamics of the dialectic and, further, to specify the manner in which the former is the concrete film form of the latter."[26] Mirroring the dialectical method, montage is seen as the resolution of conflicting shots into a new form. Preparing the film in his mind, Eisenstein writes that "in CAPITAL, for example, the themes of textile machines and machine-wreckers should collide: electric streetcar in Shanghai and thousands of coolies thereby deprived of bread, lying down on the tracks—to die."[27]

What may come as a surprise is that Marx is no less imaginative in his prose than Eisenstein is with his Shanghai streetcars and starving coolies, as a reference to Marx's "500 golden birds," or to his "rational kernel" and "mystical shell," or even to the "grotesque ideas" sprouting out of the commodity's "wooden brain," confirms.

What I would like to show, however, is that while Eisenstein shows how the aesthetic form of cinema can address discursive questions (e.g., via montage), the discourse of *Capital* itself was always already in aesthetic form, bolstered by never-ending naturalistic and vitalistic imagery (natural objects, biological processes, monsters, transmutations, and mystifications).

This vitalism in Marx heralds the dawning age of protocol, I argue, by transforming life itself into an aesthetic object. Eisenstein's goal is to transform *Capital* from a work of political economy into a *discursive event,* one that may then manifest itself in film form. Paralleling Eisenstein, I would like to give importance to "certain salient phrases" as the first step in outlining an aesthetized space within *Capital.* The moments in Marx when he lapses into metaphor and imagery appear to be his own attempt at cinematography—that is, his attempt to aestheticize the vital forms contained in the book.

---

24. Eisenstein, "Notes for a Film of *Capital,*" p. 16.

25. Eisenstein, "Notes for a Film of *Capital,*" p. 10.

26. Michelson, "Reading Eisenstein Reading *Capital,*" p. 29.

27. Eisenstein, "Notes for a Film of *Capital,*" p. 8.

The more one reads Marx, the more there is a proliferation of discourses.[28] Yes, one is the dominant language of math, science, economics, formulas, thoroughness, exhaustiveness. And another is clearly the discourse of politics—the discourse that drove Marx to break with Feuerbach and be a politico rather than simply a materialist philosopher. Another is Marx's murky analysis of the domestic sphere, both in the question of domestic labor, and in the reproduction of labor-power through sleep, food, and other nonwork activities.

Still another—what I turn to in this chapter—is the marginal language of metaphors, imagery, fantasy, poetics and suggestion. Marx's language is constantly overrun by these many levels of discourse. They mark the difference between what he is saying and why he is saying a particular thing in a particular way—from his "Force is the midwife of every old society which is pregnant with a new one,"[29] to the rhetorical comparison of "lawyer's fees, beetroot and music."[30]

Why is it important to dwell on this last discursive mode that explores capitalism's "metaphysical subtleties and theological niceties"? Because more often than not, when this mode is employed it's a question of *vitalism.* What exactly is Marx *doing* when, in the "Postface to the Second Edition" of

---

28. Several other critics have noted the existence of different discourses in Marx. For French philosopher Louis Althusser, Marx employed two distinct discourses: the ideological and the scientific. Maurice Blanchot projected a third voice in Marx, a revolutionary one called the "indirect" that is held in continuous juxtaposition with the first two (see Maurice Blanchot, "Les trois paroles de Marx," *L'Amitié* [Paris: Gallimard, 1971]). This theory of a three-way textual discordance within Marx is also championed by Jacques Derrida in his *Specters of Marx* (New York: Routledge, 1994). Both Blanchot and Derrida urge readers to embrace the multivocal nature of Marx's text. Blanchot writes that "[i]n Marx, and always coming from Marx, we see three forms of speech taking shape and finding force, all three being necessary yet separated and more-than-opposed, that is, juxtaposed" (p. 115). As Derrida writes, Blanchot asks readers "to think the 'holding together' of the disparate itself. Not to maintain together the disparate, but to put ourselves there where the disparate itself holds together, without wounding the dis-jointure, the dispersion, or the difference, without effacing the heterogeneity of the other" (p. 29).

29. Karl Marx, *Capital Volume 1,* trans. Ben Fowkes (London: Penguin, 1976), p. 916.

30. Karl Marx, *Capital Volume 3,* trans. David Fernbach (New York: Penguin, 1993), p. 953.

Capital, he advises potential critics of idealism to "discover the rational kernel within the mystical shell"?[31] Why "kernel"? Why "shell"? What is the relationship between the mystified category of the "vital form" and the fully analyzed category of the social?

Capitalism, for Marx, is *second nature.*[32] It is at once intuitive and naturalized—what Barthes would call a second-order system of signification. It is a "layer" that has been folded back on itself such that it is simultaneously its core self and its own patina. It is both raw and coded. Marx's own discourse

---

31. Marx, *Capital Volume 1,* p. 103.

32. While Marx is interested in many different vital milieus, he often fixates on nature. The most common definition Marx offers is that the natural is the *spontaneous* (see Marx, *Capital Volume 1,* pp. 201, 284, 485, 453, 471, 504, 604, 638, 645, 647, 621; Marx, *Capital Volume 3,* p. 487). Nature's spontaneity is due to its special place outside of production. Nature is a "material substratum" (Marx, *Capital Volume 1,* p. 133) that has material (but not valent) effectivity (Marx, *Capital Volume 3,* p. 955). Furthermore, nature is a type of endowment—"water, timber felled in virgin forests, and ores extracted from their veins" (Marx, *Capital Volume 1,* p. 284)—that offers forth "her gifts" (Marx, *Capital Volume 1,* p. 649) as material resources for man. In this sense nature has an "absolute fertility" (Marx, *Capital Volume 3,* p. 954) that must simply be tapped by labor power. In all of these extra-productive instances, to be natural means to be spontaneously produced. Other miscellaneous (even somewhat false) uses of nature in *Capital* include the posing of a state of nature through the Robinson Crusoe story in Book One and again in Book Three as he describes "the savage" who "must wrestle with nature to satisfy his needs" (Marx, *Capital Volume 1,* p. 959). Nature is also used in the sense of a natural biological essence to man's body, as when he claims that a man can "strike so many blows per day, walk so many steps, breathe so many breaths, produce so much work," and so on (Marx, *Capital Volume 1,* p. 366). Human nature is defined, then, in accordance with "natural needs, such as food, clothing, fuel and housing" (Marx, *Capital Volume 1,* p. 275). A concept of human nature is also seen when he describes how, when during cooperation, man "strips off the fetters of his individuality, and develops the capabilities of his species" (Marx, *Capital Volume 1,* p. 447). He also uses nature as a synonym for "in essence," as when he writes "these means of production are in and for themselves, by nature, capital" (Marx, *Capital Volume 3,* p. 963). And finally, he lets nature stand in for history, albeit an anti-essentialist history that is opposed to capitalist history, when he boldly claims that "one thing, however, is clear: nature does not produce on the one hand owners of money or commodities, and on the other hand men possessing nothing but their own labour-power. This relation has no basis in natural history" (Marx, *Capital Volume 1,* p. 273).

is no exception. And it is for this reason that Marx takes the topic of capitalism—the greatest force of naturalization known to Marx—and bathes it over and over again in the language of nature[33] and other vital forms.

I would like to begin classifying the uses of vitalistic discourse in *Capital*. While this includes biological terminology, not all biological terminology is included in my classification. There are two main exceptions—the terms *metabolic* and *organic*—where Marx employs vitalistic terminology in a strictly neutral sense. Let me consider those first.

Marx uses the term "metabolic," derived from the dynamic flow of biological processes, as an adjective to describe a relationship that is harmonious, systemic, and self-regulating, and in which skills and resources are evenly balanced yet constantly updated through a relationship of equilibrium. Marx introduces the term near the beginning of Book One of *Capital* to help describe the concept of non-owning need versus non-needing ownership: "In so far as the process of exchange transfers commodities from hands in which they are non-use-values to hands in which they are use-values, it is a process of social metabolism."[34] The systemic quality of the concept of metabolism is highlighted when Marx writes, "The exchange of commodities breaks through all the individual and local limitations of the direct exchange of products, and develops the metabolic process of human labour";[35] later Marx describes the "metabolic interaction between man and the earth"[36] in similar terms.

---

33. Nature has a very special place in Marxist science. It is its own antonym. On the one hand, the hated fetish form does nothing but *naturalize* real social relations. Yet on the other, the "natural" stands as a type of pre-capital, extra-productive force, endowing the world with useful material. The natural is also thought of as being quite literally opposed to the concept of both the social and the historical, two other crucial Marxist keywords. (The social is defined as both the supra-sensible and the supra-individual [Marx, *Capital Volume 1,* pp. 165, 207]. Social relations for Marx do not simply mean one buyer and one seller, but rather a large "network of social connections" that are "entirely beyond the control of the human agents" [Marx, *Capital Volume 1,* p. 207].)

34. Marx, *Capital Volume 1,* p. 198.

35. Marx, *Capital Volume 1,* p. 207.

36. Marx, *Capital Volume 1,* p. 637.

A distributed network is always caught, to use an expression from Deleuze and Guattari, au milieu, meaning that it is never complete, or integral to itself. The lines of a distributed network continue off the diagram. Any subsegment of a distributed network is as large and as small as its parent network. Distribution propagates through rhythm, not rebirth.

Lebbeus Woods, *Terrain 2* (1999), courtesy Henry Urbach Architecture

The contradiction at the heart of protocol is that it has to standardize in order to liberate. It has to be fascistic and unilateral in order to be utopian. It contains, as Jameson wrote of mass culture before it, both the ability to imagine an unalienated social life and a window into the dystopian realities of that life.

The NSA's Echelon station at Menwith Hill, UK
photo: Craig Stennett; courtesy the Campaign for the Accountability of American Bases

The second vitalistic term that is used quite neutrally by Marx is "organic." Like "metabolic," it is used to describe a relationship in which two or more parts fulfill individual functions for the good of the whole. This is seen in the idea of the "organic composition of capital," which simply refers to the ratio of living parts (variable capital, or labor) to dead parts (constant capital, or machines) in the total capital put in motion by the capitalist. Marx also uses "organic" to describe certain types of production, specifically as it is contrasted with the "heterogeneous" form of manufacturing in Part Four of Book One of *Capital,* or through analogy to the inner workings of a musical orchestra.[37] As in the human body, organic organization simply means division of labor into a series of "subordinate functions."[38] It is a synonym for "organization" as he describes the "organized system of machinery in the factory" as a type of "objective organism."[39] Although they are terms associated with biology, both "metabolic" and "organic" have little significance for my analysis of vitalistic imagery in Marx.

The type of vitalistic discourse seen most clearly in *Capital* is that of vital objects. Although his "rational kernel" and "mystical shell" may be the most well known, Marx is obsessed with animals, plants and minerals of all kinds. He calls the commodity form the "germ" of the money form; he uses the phrase "gold chrysalis" to describe the gold standard; he calls the money-owner a "capitalist in larval form" who will eventually emerge "as a butterfly" in the sphere of circulation; he describes the circulation of commodities as being like the flight of "500 golden birds"; he compares capitalist monetary income to the "fruits of a perennial tree," and so on.[40] The use of "congeal" and "objectification" also contributes to the proliferation of vital objects in the Marxist ontological field. When first describing the commodity, Marx writes that use-value gains value because "human labor is objectified or materialized in it."[41] Value is a "social substance," or "'objective'

---

37. See Marx, *Capital Volume 1,* pp. 461, 448–449.

38. Marx, *Capital Volume 1,* p. 644.

39. Marx, *Capital Volume 1,* p. 517.

40. For these citations, see Marx, *Capital Volume 1,* pp. 163, 227, 269; Karl Marx, *Capital Volume 2,* trans. David Fernbach (New York: Penguin, 1993), p. 495; Marx, *Capital Volume 3,* p. 960.

41. Marx, *Capital Volume 1,* p. 129.

property"[42] derived from the transmutation of human labor into the natural world.

Often Marx's vital objects take on more sinister, supernatural personalities. Specters, monsters, and vampires riddle his text. As Derrida has shown in *Specters of Marx,* the concept of haunting appears several times in *Capital* (although perhaps not as often as Derrida would lead us to believe). Marx writes twice that money "haunts" the sphere of circulation[43] through its function as universal equivalent. The interesting husband-and-wife duo Monsieur le Capital and Madame la Terre "haunt" and "bewitch"[44] the world under the influence of the Trinity Formula in Book Three. Labor itself, due to its abstract immaterial quality, is "a mere specter," claims Marx.[45] And the list continues: capital is an "animated monster"; the commodity "transcends sensuousness" and "evolves out of its wooden brain grotesque ideas," soon to evolve into a living, talking creature ("If commodities could speak . . . ," ". . . lend them his tongue"); self-valorizing value has the monstrous ability to "[bring] forth living offspring, or at least lays golden eggs"; machinery is a "vast automaton," a type of mechanical demonic monster, a lifeless machine with living appendages.[46] Yet perhaps the image of the vampire is the most sinister: "Capital is dead labour which, vampire-like, lives only by sucking living labour, and lives the more, the more labour it sucks."[47] This vampire "will not let go 'while there remains a single . . . drop of blood to be exploited.'"[48] Marx returns to the vampire theme in Book Three, describing usury in similar terms of "sucking dry" and "emasculation."[49] The theme of haunting is also evident in Marx's other work, as in the *Communist Manifesto* when he writes about the "specter haunting Europe" and the "sorcerer."[50] Yet

---

42. Marx, *Capital Volume 1,* pp. 138, 154.

43. Marx, *Capital Volume 1,* pp. 213, 226.

44. Marx, *Capital Volume 3,* p. 969.

45. Marx, *Capital Volume 3,* p. 954.

46. For these citations, see Marx, *Capital Volume 1,* pp. 302, 163, 176, 189, 255, 502, 503, 548.

47. Marx, *Capital Volume 1,* p. 342.

48. Marx, *Capital Volume 1,* p. 416.

49. Marx, *Capital Volume 3,* p. 731.

50. Karl Marx and Friedrich Engels, *The Communist Manifesto* (New York: Signet, 1998), pp. 49, 57.

perhaps the epitome of Marx's sinister monsters may be seen in a more reserved but prescient description of the type of "artificial life" that is postulated to explain the workings of individuals within a totality: "Each individual capital forms only a fraction of the total social capital, a fraction that has acquired independence and been endowed with individual life, so to speak."[51] But why does he stop short by saying "so to speak"? Is this an indication of Marx's anxiety about self-aestheticization?

A third category (after the world of objects and the world of supernatural objects) within Marx's vitalist discourse is that of natural processes. As in the other categories, the text is overflowing with rich examples. Natural process imagery includes "fermentation," "crystalization," "precipitation," and plain old "alchemy" (by which simple commodity objects are transmutated into "the money crystal").[52] Yet more fundamental to the text is the use of "congeal." Congealing is an incredibly powerful process in Marx since it involves both a change in category and a change in form. What appears on the one hand as an *activity* [category] in *humans* [form] congeals through the labor process into a *measure of value* [category] in an *object* [form]. He writes that "phantom-like," the "residue of the products of labor" congeals into "crystals" of a distinct "social substance."[53] Other naturalistic process-based imagery includes Marx's poetic allusion to the death cycle "go[ing] the way of all flesh," and the idea of consumption or extinguishment, as in the example of the labor process—"the process is extinguished in the product," and "labor uses up its material elements, its objects and its instruments. It consumes them, and is therefore a process of consumption."[54]

As mentioned earlier, the theme of mystification is introduced in the "Postface to the Second Edition" of *Capital* through Marx's example of the "rational kernel" and the "mystical shell."[55] Because "mystification" is a near synonym for "naturalization" in this text, it is clear that any discussion of

---

51. Marx, *Capital Volume 2*, p. 427.

52. For citations see Marx, *Capital Volume 1*, pp. 292, 128, 181; Marx, *Capital Volume 2*, p. 489; Marx, *Capital Volume 1*, p. 229.

53. Marx, *Capital Volume 1*, p. 128.

54. Marx, *Capital Volume 2*, p. 496; Marx, *Capital Volume 1*, pp. 287, 290.

55. Marx, *Capital Volume 1*, p. 103.

mystification or hiddenness is directly bound up in the theme of nature and other vital objects. The famous Hegel headstand that Marx posits in the "Postface . . ." is based on the idea that inversion, or upside-downness, is linked directly to illusion, mystification, and misrecognition. Because Hegel (and idealists like him) was standing on his head, he couldn't *see* the world as it was. A similar epistemological inversion is at work in Marx's analysis of fetishism. For example, he writes that as the concept of labor is fetishized and thought to have value in and of itself, it is "inverted, so that it becomes its contrary."[56] Elsewhere he describes the competition theory of value as being an "upside-down" expression,[57]—even though, on the contrary, Marx prized the labor theory of value.

Hiddenness is as powerful a force of mystification (i.e., *natural*ization) as inversion is. Marx writes that truly to understand the workings of exchange, one must leave the marketplace and venture "into the hidden abode of production."[58] He also writes that the mysteries of the money form will be resolved by revealing its origins, by showing that exchange "conceals a social relation."[59] The faults of political economy are generally due to these types of illusions—for example, the "illusion that ground rent grows out of the soil, not out of society," or the "certain deception" of the "illusory character" of the uninterrogated circuit of money capital.[60] All of these examples point to a single idea in Marx, what he calls the social hieroglyphic. The social hieroglyphic refers to something that does not announce on its surface what it *is* on the inside. He writes that "value, therefore, does not have its description branded on its forehead; it rather transforms every product of labour into a social hieroglyphic."[61] The social hieroglyphic is a characteristically "natural" object for Marx; it must be *denaturalized,* demystified. Yet the mystification of capital into natural processes is elaborate: "The further we trace out the valorization process of capital, the more is the capital relationship

---

56. Marx, *Capital Volume 1,* p. 677.

57. Marx, *Capital Volume 3,* p. 331.

58. Marx, *Capital Volume 1,* p. 279.

59. Marx, *Capital Volume 1,* pp. 139, 149.

60. Marx, *Capital Volume 1,* p. 176; Marx, *Capital Volume 2,* p. 141.

61. Marx, *Capital Volume 1,* p. 167.

mystified and the less are the secrets of its internal organization laid bare."[62] He describes capitalist analyses of the circuits of capital as "misrepresenta-tion[s]" and those of surplus value as "mislead[ing]" the capitalist.[63] The Trinity Formula at the end of Book Three "completes the mystification of the capitalist mode of production"[64] by creating false links among the compo-nents of wealth and their sources.

These concepts of mystification and naturalization may be defined fur-ther, both more generally in the concept of "form of appearance" and more specifically in the fetish form.[65] Form of appearance refers to a typically Marx-

---

62. Marx, *Capital Volume 3*, p. 139.

63. Marx, *Capital Volume 3*, pp. 516, 236.

64. Marx, *Capital Volume 3*, p. 969.

65. Fetishism is the unfortunate epistemological outcome of the form of appearance. Fetishism is how the form of appearance *is thought through*. Fetishism is a failure to read the form; it is a failure to note that the form of appearance has taken place; it is the tendency to mistake the form of appearance for the real thing. Since de*natural*ization is Marx's primary tool to combat fetishism, fetishism itself must be thought of as a type of *naturalization*. Nature runs parallel to fetish. Situated at the theoretical heart of *Capital,* the fetish has a special function for Marx. As *Capital* progresses so does the definition of the fetish. In Book One, readers are introduced to the fetish via grotesque, supernaturalistic imagery, yet by the end of Book Three the fetish has flowered into its "consummate form" (Marx, *Capital Volume 3,* p. 523), what Marx calls the automatic fetish.

Derrida has commented interestingly on the relationship between the fetish and the specter, shown previously to be integral to the category of (super)naturalistic monsters and ghosts. The spectral sits in somewhat of an antagonistic relation with the fetish form. On the money form, he writes that "Marx always described money, and more precisely the monetary sign, in the figure of appearance or simulacrum, more exactly of the ghost" (Derrida, *Specters of Marx,* p. 45). Like money, the specter has the ability to summon a remainder (a revenant, or a shadow). Derrida even hints at the commodity fetish's ability to gaze (in a way similar to some of Benjamin's observations). Derrida notes that "this spectral someone other looks at us [con-cerns us], we feel ourselves being looked at by it, outside of any synchrony, even before and be-yond any look on our part" (p. 7). The fetish is figured in much the same way as the specter. Like the relationship of appearance to disappearance in the commodity fetish (i.e., value as the form of appearance of use-value, consumption as the using up, disappearance, or realization of value, etc.), the specter too has a unique position between disappearance and reappearance—

ian dialectical process in which objects or qualities transform into their opposites through the process of representation. In its simplest usage, form of appearance means representation, or a "mode of expression."[66] For example, Marx writes that "exchange-value [is] the necessary mode of expression, or form of appearance, of value," and later he writes that "use-value becomes the form of appearance of its opposite, value."[67] Both use-value and exchange-value are negated and expressed as their mutual opposite, value. Form of appearance can mean simple "mirroring," or it can mean a more poetic, allegorical transformation as in the case of "the sheep-like nature of the Christian" transforming into the "Lamb of God."[68] Marx writes that "a change of form must occur"—that is, a "metamorphosis"—before commodities are able to be sold, "capital is always involved in this movement of transition, this metamorphosis of form."[69] Fundamentally, it is a changing of form via some type of imaginary or illusory representation, like the waterfall used for power that can have no "price" except through an "irrational expression concealing a real economic relationship."[70] Finally, form of appearance in its most advanced usage gestures toward what one would today call a theory of ideology: "What is true of all forms of appearance and their hidden background is also true of the form of appearance 'value and price of labour', of 'wages', as contrasted with the essential relation manifested in it, namely the value and price of labor power. The forms of appearance are reproduced directly and *spontaneously,* as current and usual *modes of thought.*"[71] "Spontaneity"—a near synonym for "natural" in Marx—combines with the

---

it is an economy. In relation to the specter, Derrida writes that "there is something disappeared, departed in the apparition itself as reapparition of the departed" (p. 6). In a manner unappealing to vulgar Marxists, Derrida in fact *privileges* the fetish form, since it is more helpful to deconstruction intact than demystified.

66. Marx, *Capital Volume 1,* p. 127.

67. Marx, *Capital Volume 1,* pp. 128, 148.

68. Marx, *Capital Volume 1,* pp. 144, 143.

69. Marx, *Capital Volume 1,* pp. 203, 260; Marx, *Capital Volume 3,* p. 379.

70. Marx, *Capital Volume 3,* p. 787.

71. Marx, *Capital Volume 1,* p. 682, emphasis mine.

misrecognition inherent in the wage system to produce an affect (the "mode of thought"). To simplify the formula: natural misrecognition = ideology. For this reason I argue that form of appearance is an incredibly powerful moment in Marx's theorization of vital forms.

After all of its vital objects and processes, supernatural monsters, transmutations and mystifications—after all of this, is *Capital* still simply a work of economics? Or is it a much broader analysis, one that reveals how modern life is an aesthetic space through and through? As Eisenstein was able to see, and I hope the analysis in this chapter also shows, *Capital* is an aesthetic object. The confluence of different discourses in *Capital,* both vitalistic and economic, proves this. The use of vitalistic imagery, no matter how marginalized within the text, quite literally *aestheticizes capitalism.* It turns capitalism into media. Perhaps then the conventional wisdom on *Capital,* that Marx's goal was to denaturalize the apparatus of capitalism, can be rethought. The existence in the text of vital forms allows for both an intuitive and estranged capitalistic apparatus.

Emergence of Artificial Life Forms (Matter Becoming Life)

Man has lost his *soul;* in return, however, he gains his *body.*
—GEORG VON LUKÁCS, "Thoughts Toward an Aesthetic of the Cinema"

The vital quality of pure matter has long haunted the modern era. Be it the monster in Shelley's *Frankenstein,* the commodity in Marx's *Capital,* or the murdering robot in Čapek's *R.U.R.,* the emergence of autonomous vital forms appears as a distinct trend in the last two hundred years of contemplative thought.

Much work has been done on this subject in the field of epistemology and cognitive science. During a 1959 meeting organized by the New York University Institute of Philosophy entitled "The Dimensions of Mind," Norbert Wiener and others pondered the epistemological condition of mind in the context of the machine. Later, writers such as Marvin Minsky and Daniel Dennett have considered the theoretical possibilities and limits of computerized thought.

Several theories of life are at play in this intellectual milieu. In what might be dubbed the "computers can never do what our brains can do" ideology, Hubert Dreyfus argues that there are theoretical limits to any type of artificial intelligence.[72] In a similar vein, Leopoldseder recounts that "[i]n a personal interview, the biophysician and cybernetics researcher Heinz von Foerster—one of the fathers of constructivism—answered the question of whether there is a relation between the human brain and the computer with a 'yes and no.' Then he added: 'No computer is a brain, but all brains are computers.'"[73] This debate is ongoing, with one camp claiming that computers will someday perfectly model the workings of the human brain and the other camp claiming that the *je ne sais quoi* of human thought is fundamentally different than the hard, rigid world of computer code.

Yet this book has very little to say about questions epistemological. Protocol is not a theory of mind. Nor, following Jonathan Crary and Sanford Kwinter in their forward to the fascinating collection *Incorporations,* is protocol a theory of the body. For, as they write, "Our topic is the problem of <u>life</u> itself, understood as a complex, labile, overtone structure, neither dependent upon, nor reducible to, an organic substrate or historical object—in short, to what contemporary habit too knowingly calls 'the body.'"[74] Instead protocological life is considered here as "the forces—aesthetic, technical, political, sexual—with which things combine in order to form novel aggregates of pattern and behavior."[75] Indeed, protocol is a theory of the confluence of life and matter (and ultimately we will see that protocol shows how life *is* matter).

As an introduction to the emergence of autonomous life forms in the material realm, let me first consider the theory of life that is known as the "anti-entropic" position. The anti-entropic position states, simply, that life is precisely that force *that resists entropy.* Entropy is the physical principle derived from thermodynamics that states that, in any given system, things will tend to "fall apart" or tend toward disorder. Moreover, entropy means that

---

72. This might also be dubbed the "computers don't know shit" ideology. See Stewart Brand, "SPACEWAR: Fanatic Life and Symbolic Death among the Computer Bums," *Rolling Stone,* December 7, 1972, p. 58.

73. Leopoldseder, "Forward," p. 6.

74. Jonathan Crary and Sanford Kwinter, "Forward," *Incorporations* (New York: Zone, 1992), p. 13.

75. Crary and Kwinter, "Forward," p. 13.

information, defined as any nonrandom measurement or quality, has a tendency to be forgotten. This physical principle is seen throughout nature: When something falls apart, the information about its organization is in essence forgotten and chaotic arrangement sets in in the form of decay.

Since living animals tend to resist the entropic force, they are considered to violate this law of thermodynamics. Living animals don't fall apart during their life spans, and furthermore they resist informational disorder after death by propagating genetic information to their progeny. This defines their identity as vital forms. Genes and memes (themselves also genes, but in the realm of culture) are two ways of moving against the entropic force, by conserving information, even augmenting it, from living being to living being.

Flusser paraphrases this position well:

Like no other known living creature, we pass on not only inherited but also acquired information to future generations. In doing so, we negate nature twice: the second law of thermodynamics states that in nature all information has a propensity to be forgotten. Living creatures negate this principle, since they store and pass on genetic information [genes]. And Mendel's law states that acquired information cannot be transmitted from one organism to the next. Our species defies this law, too, for it stores acquired information [memes] in a cultural memory accessible to successive generations.[76]

Thus, living beings are characterized as anti-entropic, vitality being the opposite of entropy, the capacity to resist entropy.

Peter Weibel extends this position by claiming that living forms are defined in terms of their "viability," their ability to function independently in the world: He defines a living organism as "a system characterized by its propensity to react relatively independently to any number of inputs . . . [V]iability denotes the possession of lifelike properties with the development of lifelike behavior."[77]

---

76. Flusser, "Memories," p. 202.

77. Peter Weibel, "The World as Interface: Toward the Construction of Context-Controlled Event-Worlds," in *Electronic Culture,* ed. Timothy Druckrey (New York: Aperture, 1996), pp. 348, 349.

Deleuze also agrees with this position by posing the rhetorical question "Is not life this capacity to resist force?"[78]—a capacity borne from the power of life forms to resist the imposition of biopolitical power, a type of macro-vital control that I've already discussed in more detail.

The mathematician Norbert Wiener states this position interestingly when he writes about the ability of life itself to resist the decaying forces of the universe at large:

As entropy increases, the universe, and all closed systems in the universe, tend naturally to deteriorate and lose their distinctiveness . . . But while the universe as a whole, if indeed there is a whole universe, tends to run down, there are local enclaves whose direction seems opposed to that of the universe at large and in which there is a limited and temporary tendency for organization to increase. *Life finds its home in some of these enclaves.*[79]

Wiener's position is, thus, what one might today call Deleuzian. Wiener sees entropy as a gradual procession toward the Plane of Immanence, Deleuze's term for the undifferentiated, contingent state matter finds itself in when it has yet to organize itself in any identifiable way. Life, then, is a type of stratification within that Plane. It is, quite literally, an *organ*ization, a set of "patterns that perpetuate themselves."[80]

What makes Wiener's theory so radical, however, is that he recognized that machines also resist entropy. "The machine, like the living organism, is," Wiener writes, "a device which locally and temporarily seems to resist the general tendency for the increase of entropy. . . . it can produce around it a local zone of organization in a world whose general tendency is to run down."[81]

---

78. Deleuze, *Foucault,* p. 93.

79. Norbert Wiener, *The Human Use of Human Beings: Cybernetics and Society* (New York: Da Capo, 1950), p. 12, emphasis mine. Wiener also describes life as a "local enclave" resisting environmental decay on page 95.

80. Wiener, *The Human Use of Human Beings,* p. 96.

81. Wiener, *The Human Use of Human Beings,* p. 34. Charles Petzold writes, as do computer hardware scientists at large, that the capacity for machines to preserve information derives from the "flip-flop" switch, a type of logical circuit that saves state from moment to moment:

It's not simply that machines are like people, or that people are like machines, but that *both entities are like something else,* what Wiener calls "communicative organisms," or what today might be called "information organisms." These are the same organisms that live inside protocol.

The gulf separating matter and life was further closed by Wiener with his theory of dynamic systems, known as cybernetics. The theory of cybernetics began with the simple idea of *feedback,* which he describes as "the property of being able to adjust future conduct by past performance."[82]

Yet beyond this Wiener recognized that, due to what he saw as a special isomorphism between electronic computers and the human nervous system, it would be very easy to insert "artificial sense organs" as feedback mechanisms,[83] rather than relying on one's eyes and ears. Thus, it is not out of some science fiction urge that Wiener merges man and machine, but simply that if one views the world in terms of information (or in Wiener's vocabulary, in terms of "command" and "control"—but these words are nearly synonymous with information for my purposes), then there is little instrumental difference between man and machine since both are able to affect dynamic systems via feedback loops. In this way the cybernetic system of man and machine is born.[84] Its virtues are balance, self-regulation, circularity, and control. In a word, protocol.

"[a] flip-flop circuit *retains information.* It 'remembers.' . . . A flip-flop is very much like a seesaw. A seesaw has two stable states, never staying long in that precarious middle position. . . . flip-flops are essential tools. They add memory to a circuit to give it a history of what's gone on before." See Petzold, *Code,* p. 161.

82. Wiener, *The Human Use of Human Beings,* p. 33. Wiener also writes that the "control of a machine on the basis of its *actual* performance rather than its *expected* performance is known as *feedback*" (p. 24).

83. Wiener, *Cybernetics,* p. 26.

84. Many others have followed in Wiener's footsteps. In 1960 J. C. R. Licklider, an early theorist and researcher of computer networks, wrote about what he called the "man-machine symbiosis." Marshall McLuhan also claimed that technology itself is nothing but an extension of man's nervous system. Computer pioneer Douglas Engelbart considered technology to be simply an augmentation of the human faculties. For relevant texts by Licklider and Engelbart, see Randall Packer and Ken Jordan, *Multimedia: From Wagner to Virtual Reality* (New York: Norton, 2001). Other theorists such as Donna Haraway have quite literally fused human and

Wiener's prose is tinged by anxiety over what he considered to be the vast potential for his scientific work to contribute to the "concentration of power . . . in the hands of the most unscrupulous."[85] Writing in the shadow of World War II and the atomic bomb, Wiener exhibits a grave concern, not only with the bomb but also with more general social exploitation, be it in the form of a recently defeated Nazism or a once more bullish American capitalism (he does not tell readers which). He does say that upon formulating his theory of cybernetics, his first instinct was to warn organized labor, noting: "I did manage to make contact with one or two persons high up in the C.I.O., and from them I received a very intelligent and sympathetic hearing."[86]

Yet I would argue that the proximity between man and machine in late-twentieth-century life has more utopian possibilities in Wiener's thought than it has derisive possibilities. Wiener is important because he valued the contingency of matter, be it man or machine. He recognized that material reality is the *most* important thing and that, contrary to a more static Newtonian view of matter, it *can change.*[87] The self-determinism of material systems is therefore the essence of cybernetics, and it is a positive essence, one that also reflects the positive potential of protocological organization.

Artificial Life

I personally believe that reproducing programs are living beings in the information environment.

—FREDERICK COHEN, *A Short Course on Computer Viruses*

Pseudo-artificial entities such as robots have been in existence for many years already.[88] The emergence of "artificial life" proper happens as computers shift

---

machine into what she calls the cyborg. See Donna Haraway, *Simians, Cyborgs and Women* (New York: Routledge, 1991).

85. Haraway, *Simians, Cyborgs and Women,* p. 29.

86. Haraway, *Simians, Cyborgs and Women,* p. 28.

87. For more on the "contingency of events" see Wiener, *The Human Use of Human Beings,* p. 8. Wiener's interest in "possibility" and "distribution," inspired by Gibbs's work connecting the fields of physics and statistics, is also proto-protocological (pp. 12, 8).

88. The coinage "robot" is attributed to the writer Karel Čapek and derives from a Czech word meaning "serf." For more on robots and other automata, see Julien Offray de la Mettrie,

from being primarily linear calculation machines to being clusters of parallel, distributed submachines.

In computer science, this shift is characterized by the change from "procedural" (or linear) programming to so-called object-oriented programming. In procedural programming, one inputs data and then operates on that data in a linear manner. Loops may occur, but in general a series of commands are read, interpreted, and executed as a consecutive chain of events. Object-oriented programming, on the other hand, treats all code as a series of simultaneously generated entities, with each entity possessing its own qualities and actions. Object-oriented programming was first popularized by the C++ programming language.[89] Since then nearly all of today's programming languages are able to function in such a capacity. The Java programming language, created by Sun Microsystems, is entirely object-oriented.

It is possible to map this shift historically. Sherry Turkle writes that this shift—from procedural to object-oriented—follows the shift from the modern to the postmodern eras. In what she calls "the modernist computational aesthetic" the computer was viewed as a "giant calculator," and programming it "was a cut-and-dried technical activity whose rules were crystal clear."[90] However, in today's multifaceted, distributed environments, "computational models of the mind often embrace a postmodern aesthetic of complexity and decentering. Mainstream computer researchers no longer aspire to program intelligence into computers but expect intelligence to emerge from the interactions of small subprograms."[91] This shift, from centralized

---

Machine Man and Other Writings (London: Cambridge University Press, 1996); Villiers de L'Isle Adam, *L'Eve future* (Paris: Fasquelle, 1921); Raymond Ballour, "Ideal Hadaly," *Camera Obscura* no. 15 (Fall 1986); Annette Michelson, "On the Eve of the Future: The Reasonable Facsimile and the Philosophical Toy," *October* 29 (1984); Fritz Lang's *Metropolis* and the False Maria; Issac Asimov's three rules for robots; Jasia Reichardt, *Robots* (London: Thames & Hudson, 1978); John Cohn, *Human Robots in Myth and Science* (London: Allen & Unwin, 1976).

89. C++ (pronounced "see-plus-plus") was developed by Bjarne Stroustrup in 1980. Stroustrup's language was built on the preexisting C language written a decade earlier by Brian Kernighan and Dennis Ritchie (who had previously worked on a lesser version called B).

90. Sherry Turkle, "Who Am We?," *Wired,* January 1996, p. 149.

91. Turkle, "Who Am We?," p. 149.

procedural code to distributed object-oriented code, is the most important shift historically for the emergence of artificial life.

In one of the most celebrated instances of artificial life research, the computer scientist Tom Ray created a virtual world called *Tierra* in which living "digital organisms" self-replicate and evolve according to Darwinian rules.[92] Ray writes:

Life on Earth is the product of evolution by natural selection operating in the medium of carbon chemistry. However, in theory, the process of evolution is neither limited to occurring [sic] on the Earth, nor in carbon chemistry. Just as it may occur on other planets, it may also operate in other media, such as the medium of digital computation. . . .

The Tierra C source code creates a virtual computer and its Darwinian operating system, whose architecture has been designed in such a way that the executable machine codes are evolvable. This means that the machine code can be mutated (by flipping bits at random) or recombined (by swapping segments of code between algorithms), and the resulting code remains functional enough of the time for natural (or presumably artificial) selection to be able to improve the code over time.[93]

The natural resources in *Tierra* are not food and shelter but CPU time and memory space. Evolution appears in fast motion as Ray's digital organisms split, recombine, and mutate as if they were the organic code of DNA.

The ecologically minded Ray has gone so far as to advocate the institution on the Net of a wildlife preserve for digital organisms. He writes: "I proposed to create a very large, complex and inter-connected region of cyberspace that will be inoculated with digital organisms which will be allowed to evolve freely through natural selection"[94]—the goal of which is to model the

---

92. For other examples of artificial life computer systems, see Craig Reynolds's "boids" and the flocking algorithm that governs their behavior, Larry Yaeger's "Polyworld," Myron Krüger's "Critter," John Conway's "Game of Life," and others.

93. Tom Ray, "What Tierra Is," available online at http://www.hip.atr.co.jp/~ray/tierra/whatis.html.

94. Tom Ray, "Beyond Tierra: Towards the Digital Wildlife Reserve," available online at http://www1.univap.br/~pedrob/PAPERS/FSP_96/APRIL_07/tom_ray/node5.html.

spontaneous emergence of biodiversity, a condition believed by many scientists to be the true state of distribution of genetic information in a Nature that is unencumbered by human intervention.

Yet in the end, the surprise is not really that *Tierra* acts like the world of living beings, but instead that the world of living beings acts like the computerized space of *Tierra*—this is the ontological rub guiding the emergence of nonorganic life.

Life as Medium (Life Becoming Matter)

I assert that, further to the anti-entropic theory of life (which by itself has little to say about protocol), life forms, both artificial and organic, exist in any space where material forces are *actively aestheticized,* resulting in a type of sculpted materiality, a materiality in which vital agents are managed, organized, affected, and otherwise made aesthetically active. (Thus was the goal of my earlier analysis of Marx.) For my purposes, the same protocological forces that regulate data flows within contingent environments such as distributed networks are the same forces that regulate matter itself.

In his introduction to Branden Hookway's intriguing pamphlet *Pandemonium: The Rise of Predatory Locales in the Postwar Period,* Kwinter suggests that the nineteenth century saw the emergence of a new phenomenon of "distributed management" appearing at the intersection of materialist philosophy and the science of thermodynamics. This new form of distributed management condenses into the form of the *demon,* an autonomous agent that is immanent to a material milieu:

The demons of *Pandemonium* descend primarily from the demons of James Clerk-Maxwell's nineteenth-century thought experiments with heat and gases, and the freshly discovered rule-based behaviors that determine them. It might be said that materialism found far deeper and more reliable apologists in the nineteenth-century thermodynamicists than it did in Marx or Engels . . . Marx saw social life as buffeted, driven and shaped by irresistible but logically ordered economic forces whose end product was assumed to be inevitable: the "catastrophic" leap to democratic equilibrium. Heat scientists likewise formulated the mysterious "second law": all closed systems lose order, all decay toward entropy. Yet for neither Marx nor Clerk-Maxwell was this simply a mechanical or energetic relation; something new had clearly emerged, *a system of distributed management that could not be dissociated from the material milieus whose particles and behavioral quirks were being managed.* This was the birth of information science.[95]

The "information age"—a term irreverently tossed to and fro by many critics of contemporary life—is not simply that moment when computers come to dominate, but is instead that moment in history when matter itself is understood in terms of information or code. At this historical moment, protocol becomes a controlling force in social life.

Kittler documents this passage well when he writes about the transformation from a "kingdom of sense" in the year 1800 to a "kingdom of pattern" in 1900.[96] But what has been overlooked is that the transformation of matter into code is not only a passage from the qualitative to the quantitative *but also a passage from the non-aesthetic to the aesthetic—the passage from non-media to media.*

So, in 1953, when Watson and Crick discovered DNA (perhaps not anticipating its subsequent ascendance to the supreme definition of life itself fifty years later), they prove not simply that life is an informatic object (that much had been clear for decades with the Bertillon system of criminal phrenology, the quantification of human movement by Muybridge, and so on) but rather that life is an aesthetic object; it is a *double helix,* an elegant, hyper-Platonic form that rises like a ladder into the heights of aesthetic purity.

Life was no longer a "pair of ragged claws / Scuttling across the floors of silent seas" (Eliot), it was a code borne from pure mathematics, an object of aesthetic beauty, a double helix! This historical moment—when life is defined no longer as essence, but as code—is the moment when life *becomes a medium.*[97]

---

95. Sanford Kwinter, "Introduction: War in Peace," in Branden Hookway, *Pandemonium: The Rise of Predatory Locales in the Postwar Period* (Princeton: Princeton Architectural Press, 1999), pp. 9–10, emphasis mine.

96. See Friedrich Kittler, *Discourse Networks, 1800/1900,* trans. Michael Metteer and Chris Cullens (Stanford: Stanford University Press, 1990).

97. Many contemporary artists consider life to be a medium in quite literal ways. In particular, see the work of Eduardo Kac, CAE, Biotech Hobbyist Magazine (Natalie Jeremijenko and Heath Bunting), Orlan, and Stelarc.

This historical shift was demonstrated in the winter of 1995 when Leonard Adleman of the University of Southern California created a new computer called the TT-100. What was unusual about this new computer was not simply that it eschewed Microsoft's recently released Windows 95 operating system, but that it required *no* operating system at all. Adleman's computer was a DNA computer, an organic computation machine contained in less than a drop of fluid.

DNA computing works by translating the binary code of ones and zeros into the nonbinary but equally quantitative code of DNA. Each strand of DNA may be correlated with a computational problem, and since millions of strands of DNA can be amassed in a relatively small space, exposing these strands to chemical reactions causes an exponentially large number of computations to occur at the same moment.

As one journalist put it, the advantage of DNA computing is that "chemical reactions occur very fast and in parallel, so that if the DNA molecules are synthesized with a chemical structure that represents numerical information, a vast amount of number-crunching is done as the reaction proceeds."[98]

In Adleman's machine *life had become a medium.* At that moment, life itself formed the building blocks of the basic calculations of computer science.

This moment was further demonstrated in the summer of 1995 when the Max Planck Institute announced its success in creating a two-way communication link between a living neuron and a silicon chip. As *The New York Times* wrote, "The accomplishment announced Monday [August 21] has thus established a signaling channel between a nerve cell and a silicon chip that works in both directions."[99] The connection between human neuron and silicon chip was a physical reinforcement of a larger trend linking organic and artificial life, a trend perhaps most associated with the name Donna Haraway who as early as 1984 had written about the "hybrid of machine and organism."[100]

---

98. Gina Kolata, "A Vat of DNA May Become Fast Computer Of the Future," *The New York Times,* April 11, 1995, p. C1.

99. *The New York Times,* cited in Timothy Druckrey, "Wild Nature, Free Radicals, and The Nerve Sell . . . ," DEAF95 Symposium, November 24, 1995, available online at http://www.v2.nl/DEAF/persona/druckrey-txt.html.

100. See Donna Haraway, "A Cyborg Manifesto: Science, Technology, and Socialist-Feminism in the Late Twentieth Century," *Simians, Cyborgs, and Women* (New York: Routledge, 1991), p. 149.

Yet this shift is much older than that. Malraux's concept of the "imaginary museum" showed how museums aestheticize material objects twice over, creating a social space within the institution that is as thick with its own affective power as any artwork within its walls.

The Zone Improvement Plan (ZIP), created by Robert Moon in 1963, helped convert the older letter-based semantics of postal addresses into new number-based zip codes (not unlike the Internet's IP addresses). This move, plus the concurrent shift in the telephone system from word-based exchanges to the more common seven-digit telephone *number,* helped usher in a new era of informatization of personal lives. One's lived experience was no longer tied to material realities, but instead was understood in terms of numbers—a telephone number, a zip code, a social security number, an IP address, and so on.

The science of measuring the human body and deriving digital signatures from it is called biometrics. What used to stand for identity—external objects like an ID card or key, or social relations like a handshake or an interpersonal relationship, or an intangible like a password that is memorized or digitized—has been replaced in recent decades by biometric examinations such as identity checks through eye scans, blood tests, fingerprinting, etc.

Criticism of biometrics has thus far focused largely on privacy, for physical traits are considered to be so intimately connected to one's identity that the measuring and tracking of them is an infringement on privacy.

However, I would like to point out instead that biometrics does something much more important. It considers living human bodies not in their immaterial essences, or souls, or what have you, but in terms of quantifiable, recordable, enumerable, and encodable characteristics. It considers life as an aesthetic object. It is the natural evolution of Marx's theory of second nature discussed earlier.

Biometrics is important, therefore, not because it infringes on privacy, but because it has redefined what counts as proof of the true identity of material life forms. Authenticity (identity) is once again inside the body-object, yet it appears now in sequences, samples, and scans.

This quantification of living forms has reached an interesting hypertrophy in the computer technique of "collaborative filtering." Collaborative filtering, also called suggestive filtering and included in the growing field of "intelligent agents," allows one to predict new characteristics (particularly one's so-called desires) based on survey data. A user answers a series of questions

Table 3.1

Control Matrix

Era	Machine	Energy mode	Disciplinary mode	Control diagram	Virtue
Feudal/early modern	Pendula, levers, pulleys	Slave, vassal	Tithe, feudal, tribute	Violence	Loyalty, allegiance, secrecy
Modern	Oil, steam, nuclear	Wage labor	Efficiency	Bureaucracy	Presence, centralization, efficiency
Postmodern, Empire	Computers	Immaterial labor, information	Debugging	Protocol	Fidelity, pattern, algorithm
The future	Bioinformatics	Life	Therapy	Physics	Openness

about his or her likes and dislikes, thus setting up a personal "profile." The profile is entered into a pool of other profiles supplied by other users. Then, statistical algorithms suggest other likes and dislikes of the user, based on the similarity of his or her profile to other users' profiles in the pool.

As a representative of industry pioneer (and Microsoft casualty) Firefly described in email correspondence: "A user's ratings are compared to a database full of other members' ratings. A search is done for the users that rated selections the same way as this user, and then the filter will use the other ratings of this group to build a profile of that person's tastes."

What makes this technique so different from other survey-based predictive techniques is the use of powerful algorithms to determine and at the same time inflect the identity of the user. However because collaborative filtering works through a process of interpellation (selecting data *interior* rather than exterior to a given set), no improvement in the overall data pool is possible. Thus, collaborative filtering ensures structural homogeneity rather than heterogeneity. While any given user may experience a broadening of his or her personal tastes, the pool at large becomes less and less internally diverse.

Collaborative filtering is therefore an extreme example of the protocological organization of real human people. Personal identity is formed only on certain hegemonic patterns. In this massive algorithmic collaboration, the user is always suggested to be like someone else, who, in order for the system to work, is already like the user to begin with! Collaborative filtering is a

Active threat (resistance)	Passive threat (delinquency)	Political mode	Stratagem	Personal crisis
Armed revolt, sedition	Betrayal, blasphemy	Revolution	Land	Abandonment
Sabotage, treason	Red tape, entropy	Disruption	Territory, deterrence	Pain, emptiness
Mutation, terrorism	Randomness, noise	Disturbance	Security, containment	Terror
Irrationality	Silence, nonexistence, irresponsiveness	Hypertrophy	Peace	Contagion, eruption

synchronic logic injected into a social relation. That is to say, like the broad definition of protocol I use, collaborative filtering is a set of rules based on a pool of user dispositions that affects each member of the pool.

Part I of this book is drawing to a close. I have so far described how protocol works as a physical technology, as a formal technology, and as a political technology. By way of conclusion I offer the control matrix shown in table 3.1.

The matrix describes protocol's successes, its failures, and its future forms. Thus far I have only considered the successes. Yet in parts II and III I look at its failures and possible future forms.

II

Failures of Protocol

4

Institutionalization

In the Internet, there is no central node, and only a minimal centralized management structure, limited to a few housekeeping functions such as standards setting.
—PAUL BARAN, "Is the UHF Frequency Shortage a Self Made Problem?"

We define mechanism, not policy.
—TIM BERNERS-LEE, *Weaving the Web*

On April 12, 1994, the protocological organization of the Internet suffered a major setback. On that black Tuesday, an unsolicited commercial email message was sent systematically to each and every newsgroup in the Usenet system, violating the informational network's customary prohibition against such commercial advertisements.[1]

Spam was born. The perpetrators, Arizona lawyers Laurence Canter and Martha Seigel,[2] had effectively transformed a democratic, protocological system for exchange of ideas into a unilateral, homogenous tool for commercial solicitation.

A quick description of Usenet is as follows:

Usenet has evolved some of the best examples of decentralized control structures on the Net. There is no central authority that controls the news system. The addition of new newsgroups to the main topic hierarchy is controlled by a rigorous democratic process, using the Usenet group *news.admin* to propose and discuss the creation of new groups. After a new group is proposed and discussed for a set period of time, anyone with an email address may submit an email vote for or against the proposal. If a newsgroup vote passes, a new group message is sent and propagated through the Usenet network.[3]

This protocological covenant outlining open channels for Usenet's growth and governance, hitherto cultivated and observed by its large, diverse com-

---

Epigraphs: Paul Baran, "Is the UHF Frequency Shortage a Self Made Problem?" Paper presented at the Marconi Centennial Symposium, Bologna, Italy, June 23, 1995. Tim Berners-Lee, *Weaving the Web* (New York: HarperCollins, 1999), p. 124.

1. This standard of etiquette is articulated in Sally Hambridge and Albert Lunde's RFC on the topic, "DON'T SPEW: A Set of Guidelines for Mass Unsolicited Mailings and Postings (spam)," RFC 2635, FYI 35, June 1999. See also Sally Hambridge, "Netiquette Guidelines," RFC 1855, FYI 28, October 1995. Stopgap technical solutions for reducing the amount of spam are outlined in Gunnar Lindberg's "Anti-Spam Recommendations for SMTP MTAs," RFC 2505, BCP 30, February 1999.

2. The two document this and other questionable practices in their book *How to Make a Fortune on the Information Superhighway: Everyone's Guerrilla Guide to Marketing on the Internet and Other On-Line Services* (New York: HarperCollins, 1995).

3. Nelson Minar and Marc Hedlund, "A Network of Peers," in *Peer-to-Peer: Harnessing the Power of Disruptive Technologies,* ed. Andy Oram (Sebastopol: O'Reilly, 2001), p. 6.

munity of scientists and hobbyists, was sullied in the spam incident by the infraction of a few. The diversity of the many groups on Usenet was erased and covered by a direct-mail blanket with a thoroughness only computers can accomplish. As I stated earlier, protocol requires universal adoption. As a protocological product, Usenet is vulnerable because of this. Even a single party can exploit a weakness and, like a virus, propagate through the system with logical ferocity.

In part I I described how protocol has succeeded as a dominant principle of organization for distributed networks. Yet at the same time the spam incident of April 12, 1994, illustrates that there have been numerous instances where protocol has, in a sense, failed. The openness of the network was wrenched away from its users and funneled toward a single commercial goal. What was multiple became singular. What was contingent and detached became directed and proprietary.

Failures of protocol occur in many places of contemporary life, from the dominance of international capitalism and the World Trade Organization, itself a power center that buckled under distributed, protocological protests against it in Seattle in 1999, to the monolithic Microsoft and its battle with the U.S. Justice Department (the anti-Microsoft action is, to be precise, a failure of a failure of protocol).

By failure I mean to point out not a failure on protocol's own terms (that's what part III of this book is for), but a failure for protocol to blossom fully as a management diagram. That is to say, this section is not about how protocol doesn't work—because it does, very well—but how protocol is not allowed to work purely on its own terms.

This chapter, then, covers how protocol has emerged historically within a context of bureaucratic and institutional interests, a reality that would seem to contradict protocol. And indeed it does. (Or, as I will put it at the end of this chapter, in a sense protocol has to fail in order to succeed, to fail tactically in order to succeed strategically.) While in Paul Baran's estimation these interests are a "minimal" management structure, they have exerted influence over the network in significant ways. Proprietary or otherwise commercial interests (from the spam incident to Microsoft and everything in between) also represent a grave threat to and failure of protocol.

To date, most of the literature relating to my topic has covered protocol through these issues of law, governance, corporate control, and so on. Lawrence Lessig is an important thinker in this capacity. So I do not cover that in detail in this chapter. But in passing consider this heuristic: It is possible to think of bureaucratic interests as visiting protocol *from without* due to the imposition of a completely prior and foreign control diagram, while proprietary interests arrive *from within* as a coopting of protocol's own explosive architecture. Bureaucracy is protocol atrophied, while propriety is protocol reified. Both represent grave challenges to the effective functioning of protocol within digital computer networks.

Let me say also that this is the least significant section—and indeed because of that, the most significant—to read if one is to understand the true apparatus of protocol. The argument in this book is that bureaucratic and institutional forces (as well as proprietary interests) are together the inverse of protocol's control logic. This is why I have not yet, and will not, define protocol's power in terms of either commercial control, organizational control, juridical control, state control, or anything of the like. Protocol gains its authority from another place, from technology itself and how people program it.

To be precise, many believe that bureaucratic organizations such as ICANN (the Internet Corporation for Assigned Names and Numbers) are synonymous with protocol because they regulate and control the Net. But the opposite is true. Organizations like ICANN are the enemy of protocol because they limit the open, free development of technology. (It is for this reason that I have waited until this chapter to discuss the RFCs in detail, rather than talking about them in chapter 1.)

Likewise, the market monopoly of Intel in the field of microchips or of Microsoft in the field of personal computer software appears to many to constitute a type of protocol, a broad technical standard. But, again, market monopolies of proprietary technologies are the inverse, or enemy, of protocol, for they are imposed from without, are technically opaque, centrally controlled, deployed by commercial concerns, and so on.

As long-time RFC editor Jon Postel put it, "I think three factors contribute to the success of the Internet: (1) public documentation of the protocols, (2) free (or cheap) software for the popular machines, and (3) vendor

---

4. See Jon Postel's biographical entry in Gary Malkin's "Who's Who in the Internet: Biographies of IAB, IESG and IRSG Members," RFC 1336, FYI 9, May 1992.

independence."[4] Commercial or regulatory interests have historically tended to impinge upon Postel's three factors. Standards bodies like the Institute of Electrical and Electronics Engineers (IEEE) make a point of publishing standards that do not reference or favor any specific commercial vendor. (They accomplish this by describing how a technology should perform, not any specific design implementation, which may be linked to a specific commercial product or patented technology.) Hence, this chapter is nothing but a prophylactic. It addresses the negative influences that restrict protocol's full potential.

In short, *protocol is a type of controlling logic that operates outside institutional, governmental, and corporate power,* although it has important ties to all three.

In this day and age, technical protocols and standards are established by a self-selected oligarchy of scientists consisting largely of electrical engineers and computer specialists. Composed of a patchwork of many professional bodies, working groups, committees, and subcommittees, this technocratic elite toils away, mostly voluntarily, in an effort to hammer out solutions to advancements in technology. Many of them are university professors. Most all of them either work in industry or have some connection to it.

Like the philosophy of protocol itself, membership in this technocratic ruling class is open. "Anyone with something to contribute could come to the party,"[5] wrote one early participant. But, to be sure, because of the technical sophistication needed to participate, this loose consortium of decision makers tends to fall into a relatively homogenous social class: highly educated, altruistic, liberal-minded science professionals from modernized societies around the globe.

And sometimes not so far around the globe. Of the twenty-five or so original protocol pioneers, three of them—Vint Cerf, Jon Postel, and Steve Crocker—all came from a single high school in Los Angeles's San Fernando Valley.[6] Furthermore, during his long tenure as RFC editor, Postel was the single gatekeeper through whom all protocol RFCs passed before they could be published. Internet historians Katie Hafner and Matthew Lyon describe

---

5. Jake Feinler, "30 Years of RFCs," RFC 2555, April 7, 1999.

6. See Vint Cerf's memorial to Jon Postel's life and work in "I Remember IANA," RFC 2468, October 1998.

this group as "an ad-hocracy of intensely creative, sleep-deprived, idiosyncratic, well-meaning computer geniuses."[7]

There are few outsiders in this community. Here the specialists run the show. To put it another way, while the Internet is used daily by vast swaths of diverse communities, the standards makers at the heart of this technology are a small entrenched group of techno-elite peers. The reasons for this are largely practical. "Most users are not interested in the details of Internet protocols," Cerf observes. "They just want the system to work."[8] Or as former IETF Chair Fred Baker reminds us: "The average user doesn't write code. . . . If their needs are met, they don't especially care how they were met."[9]

So who actually writes these technical protocols, where did they come from, and how are they used in the real world? They are found in the fertile amalgamation of computers and software that constitutes the majority of servers, routers, and other Internet-enabled machines. A significant portion of these computers were, and still are, Unix-based systems. A significant portion of the software was, and still is, largely written in the C or C++ languages. All of these elements have enjoyed unique histories as protocological technologies.

The Unix operating system was developed at Bell Telephone Laboratories by Ken Thompson, Dennis Ritchie, and others beginning in 1969, and development continued into the early 1970s. After the operating system's release, the lab's parent company, AT&T, began to license and sell Unix as a commercial software product. But, for various legal reasons, the company admitted that it "had no intention of pursuing software as a business."[10] Unix was indeed sold by AT&T, but simply "as is" with no advertising, technical support, or other fanfare. This contributed to its widespread adoption

---

7. Katie Hafner and Matthew Lyon, *Where Wizards Stay Up Late: The Origins of the Internet* (New York: Touchstone, 1996), p. 145. For biographies of two dozen protocol pioneers, see Gary Malkin's "Who's Who in the Internet: Biographies of IAB, IESG and IRSG Members," RFC 1336, FYI 9, May 1992.

8. Vinton Cerf, personal correspondence, September 23, 2002.

9. Fred Baker, personal correspondence, December 12, 2002.

10. AT&T's Otis Wilson, cited in Peter Salus, *A Quarter Century of Unix* (New York: Addison-Wesley, 1994), p. 59.

by universities who found in Unix a cheap but useful operating system that could be easily experimented with, modified, and improved.

In January 1974, Unix was installed at the University of California at Berkeley. Bill Joy and others began developing a spin-off of the operating system that became known as BSD (Berkeley Software Distribution).

Unix was particularly successful because of its close connection to networking and the adoption of basic interchange standards. "Perhaps the most important contribution to the proliferation of Unix was the growth of networking,"[11] writes Unix historian Peter Salus. By the early 1980s, the TCP/IP networking suite was included in BSD Unix.

Unix was designed with openness in mind. The source code—written in C, which was also developed during 1971–1973—is easily accessible, meaning a higher degree of technical transparency.

The standardization of the C programming language began in 1983 with the establishment of an American National Standards Institute (ANSI) committee called "X3J11." The ANSI report was finished in 1989 and subsequently accepted as a standard by the international consortium ISO in 1990.[12] Starting in 1979, Bjarne Stroustrup developed C++, which added the concept of classes to the original C language. (In fact, Stroustrup's first nickname for his new language was "C with Classes.") ANSI standardized the C++ language in 1990.

C++ has been tremendously successful as a language. "The spread was world-wide from the beginning," recalled Stroustrup. "[I]t fit into more environments with less trouble than just about anything else."[13] Just like a protocol.

It is not only computers that experience standardization and mass adoption. Over the years many technologies have followed this same trajectory. The process of standards creation is, in many ways, simply the recognition

---

11. Salus, *A Quarter Century of Unix,* p. 2.

12. See Dennis Ritchie, "The Development of the C Programming Language," in *History of Programming Languages II,* ed. Thomas Bergin and Richard Gibson (New York: ACM, 1996), p. 681.

13. Bjarne Stroustrup, "Transcript of Presentation," in *History of Programming Languages II,* ed. Thomas Bergin and Richard Gibson (New York: ACM, 1996), p. 761.

of technologies that *have* experienced success in the marketplace. One example is the VHS video format developed by JVC (with Matsushita), which edged out Sony's Betamax format in the consumer video market. Betamax was considered by some to be a superior technology (an urban myth, claim some engineers) because it stored video in a higher-quality format. But the trade-off was that Betamax tapes tended to be shorter in length. In the late 1970s when VHS launched, the VHS tape allowed for up to two hours of recording time, while Betamax provided only one hour. "By mid 1979 VHS was outselling Beta by more than 2 to 1 in the US."[14] When Betamax caught up in length (to three hours), it had already lost a foothold in the market. VHS would counter Betamax by increasing to four hours and later eight.

Some have suggested that it was the pornography industry, which favored VHS over Betamax, that provided it with legions of early adopters and proved the long-term viability of the format.[15] But perhaps the most convincing argument is the one that points out JVC's economic strategy that included aggressive licensing of the VHS format to competitors. JVC's behavior is pseudo-protocological. The company licensed the technical specifications for VHS to other vendors. It also immediately established manufacturing and distribution supply chains for VHS tape manufacturing and retail sales. In the meantime Sony tried to fortify its market position by keeping Betamax to itself. As one analyst writes:

Three contingent early differences in strategy were crucial. First, Sony decided to proceed without major co-sponsors for its Betamax system, while JVC shared VHS with several major competitors. Second, the VHS consortium quickly installed a

---

14. S. J. Liebowitz and Stephen E. Margolis, "Path Dependence, Lock-In and History," *Journal of Law, Economics and Organization* 11, April 1995. Available online at http://wwwpub. utdallas.edu/~liebowit/paths.html.

15. If not VHS then the VCR in general was aided greatly by the porn industry. David Morton writes that "many industry analysts credited the sales of erotic video tapes as one of the chief factors in the VCR's early success. They took the place of adult movie theaters, but also could be purchased in areas where they were legal and viewed at home." See Morton's *A History of Electronic Entertainment since 1945,* available online at http://www.ieee.org/organizations/ history_center/research_guides/entertainment, p. 56.

large manufacturing capacity. Third, Sony opted for a more compact cassette, while JVC chose a longer playing time for VHS, which proved more important to most customers.[16]

JVC deliberately sacrificed larger profit margins by keeping prices low and licensing to competitors. This was in order to grow its market share. The rationale was that establishing a standard was the most important thing, and as JVC approached that goal, it would create a positive feedback loop that would further beat out the competition.

The VHS/Betamax story is a good example from the commercial sector of how one format can triumph over another format to become an industry standard. This example is interesting because it shows that protocological behavior (giving out your technology broadly even if it means giving it to your competitors) often wins out over proprietary behavior. The Internet protocols function in a similar way, to the degree that they have become industry standards not through a result of proprietary market forces, but due to broad open initiatives of free exchange and debate. This was not exactly the case with VHS, but the analogy is useful nevertheless.

This type of corporate squabbling over video formats has since been essentially erased from the world stage with the advent of DVD. This new format was reached through consensus from industry leaders and hence does not suffer from direct competition by any similar technology in the way that VHS and Betamax did. Such consensus characterizes the large majority of processes in place today around the world for determining technical standards.

Many of today's technical standards can be attributed to the IEEE (pronounced "eye triple e"). In 1963 IEEE was created through the merging of two professional societies. They were the American Institute of Electrical Engineers (AIEE) founded in New York on May 13, 1884 (by a group that included Thomas Edison) and the Institute of Radio Engineers (IRE) founded in 1912.[17] Today the IEEE has over 330,000 members in 150 countries. It is the world's largest professional society in any field. The IEEE works in con-

---

16. Douglas Puffert, "Path Dependence in Economic Theory." Available online at http://www.vwl.uni-muenchen.de/ls_komlos/pathe.pdf, p. 5.

17. *IEEE 2000 Annual Report,* available online at http://www.ieee.org.

junction with industry to circulate knowledge of technical advances, to recognize individual merit through the awarding of prizes, and to set technical standards for new technologies. In this sense the IEEE is the world's largest and most important protocological society.

Composed of many chapters, subgroups, and committees, the IEEE's Communications Society is perhaps the most interesting area vis-à-vis computer networking. It establishes standards in many common areas of digital communication including digital subscriber lines (DSLs) and wireless telephony.

IEEE standards often become international standards. Examples include the "802" series of standards that govern network communications protocols. These include standards for Ethernet[18] (the most common local area networking protocol in use today), Bluetooth, Wi-Fi, and others.

"The IEEE," Paul Baran observed, "has been a major factor in the development of communications technology."[19] Indeed Baran's own theories, which eventually would spawn the Internet, were published within the IEEE community even as they were published by his own employer, the Rand Corporation.

Active within the United States are the National Institute for Standardization and Technology (NIST) and ANSI. The century-old NIST, formerly known as the National Bureau of Standards, is a federal agency that develops and promotes technological standards. Because it is a federal agency and not a professional society, it has no membership per se. It is also nonregulatory, meaning that it does not enforce laws or establish mandatory standards that must be adopted. Much of its budget goes into supporting NIST research laboratories as well as various outreach programs.

ANSI, formerly called the American Standards Association, is responsible for aggregating and coordinating the standards creation process in the

18. The IEEE prefers to avoid associating its standards with trademarked, commercial, or otherwise proprietary technologies. Hence the IEEE definition eschews the word "Ethernet," which is associated with Xerox PARC where it was named. The 1985 IEEE standard for Ethernet is instead titled "IEEE 802.3 Carrier Sense Multiple Access with Collision Detection (CSMA/CD) Access Method and Physical Layer Specifications."

19. Paul Baran, Electrical Engineer, an oral history conducted in 1999 by David Hochfelder, IEEE History Center, Rutgers University, New Brunswick, NJ, USA.

United States. It is the private-sector counterpart to NIST. While it does not create any standards itself, it is a conduit for federally accredited organizations in the field who are developing technical standards. The accredited standards developers must follow certain rules designed to keep the process open and equitable for all interested parties. ANSI then verifies that the rules have been followed by the developing organization before the proposed standard is adopted.

ANSI is also responsible for articulating a national standards strategy for the United States. This strategy helps ANSI advocate in the international arena on behalf of U.S. interests. ANSI is the only organization that can approve standards as American national standards.

Many of ANSI's rules for maintaining integrity and quality in the standards development process revolve around principles of openness and transparency and hence conform with much of what I have already said about protocol. ANSI writes that

- Decisions are reached through *consensus* among those affected.
- Participation is *open* to all affected interests. . . .
- The process is *transparent*—information on the process and progress is directly available. . . .
- The process is *flexible,* allowing the use of different methodologies to meet the needs of different technology and product sectors.[20]

Besides being consensus-driven, open, transparent, and flexible, ANSI standards are also voluntary, which means that, like NIST, no one is bound by law to adopt them. Voluntary adoption in the marketplace is the ultimate test of a standard. Standards may disappear in the advent of a new superior technology or simply with the passage of time. Voluntary standards have many advantages. By not forcing industry to implement the standard, the burden of success lies in the marketplace. And in fact, proven success in the marketplace generally predates the creation of a standard. The behavior is emergent, not imposed.

---

20. ANSI, "National Standards Strategy for the United States." Available online at http://www.ansi.org, emphasis in original.

On the international stage several other standards bodies become important. The International Telecommunication Union (ITU) focuses on radio and telecommunications, including voice telephony, communications satellites, data networks, television, and, in the old days, the telegraph. Established in 1865, it is the world's oldest international organization.

The International Electrotechnical Commission (IEC) prepares and publishes international standards in the area of electrical technologies including magnetics, electronics, and energy production. They cover everything from screw threads to quality management systems. IEC is comprised of national committees. (The national committee representing the United States is administered by ANSI.)

Another important international organization is ISO, also known as the International Organization for Standardization.[21] Like IEC, ISO grew out of the electro-technical field and was formed after World War II to "facilitate the international coordination and unification of industrial standards."[22] Based in Geneva, but a federation of over 140 national standards bodies including the American ANSI and the British Standards Institution (BSI), its goal is to establish vendor-neutral technical standards. Like the other international bodies, standards adopted by the ISO are recognized worldwide.

Also like other standards bodies, ISO develops standards through a process of consensus-building. Its standards are based on voluntary participation, and thus the adoption of ISO standards is driven largely by market forces (as opposed to mandatory standards that are implemented in response to a governmental regulatory mandate). Once established, ISO standards can have massive market penetration. For example, the ISO standard for film speed (100, 200, 400, etc.) is used globally by millions of consumers.

Another ISO standard of far-reaching importance is the Open Systems Interconnection (OSI) Reference Model. Developed in 1978, the OSI Reference Model is a technique for classifying all networking activity into seven

---

21. The name ISO is in fact not an acronym, but derives from a Greek word for "equal." This way it avoids the problem of translating the organization's name into different languages, which would produce different acronyms. The name ISO, then, is a type of semantic standard in itself.

22. See http://www.iso.ch for more history of the ISO.

abstract layers. Each layer describes a different segment of the technology behind networked communication, as described in chapter 1.

Layer 7   Application
Layer 6   Presentation
Layer 5   Session
Layer 4   Transport
Layer 3   Network
Layer 2   Data link
Layer 1   Physical

This classification, which helps organize the process of standardization into distinct areas of activity, is relied on heavily by those creating data networking standards.

In 1987 ISO and IEC recognized that some of their efforts were beginning to overlap. They decided to establish an institutional framework to help coordinate their efforts and formed a joint committee to deal with information technology called the Joint Technical Committee 1 (JTC 1). ISO and IEC both participate in the JTC 1, as well as liaisons from Internet-oriented consortia such as the IETF. ITU members, IEEE members, and others from other standards bodies also participate here. Individuals may sit on several committees in several different standards bodies, or simply attend as ex officio members, to increase inter-organizational communication and reduce redundant initiatives between the various standards bodies. JTC 1 committees focus on everything from office equipment to computer graphics. One of the newest committees is devoted to biometrics.

ISO, ANSI, IEEE, and all the other standards bodies are well-established organizations with long histories and formidable bureaucracies. The Internet, on the other hand, has long been skeptical of such formalities and spawned a more ragtag, shoot-from-the-hip attitude about standard creation.[23] I fo-

---

23. The IETF takes pride in having such an ethos. Jeanette Hofmann writes: "The IETF has traditionally understood itself as an elite in the technical development of communication networks. Gestures of superiority and a dim view of other standardisation committees are matched by unmistakable impatience with incompetence in their own ranks." See Hofmann, "Govern-

Chapter 4

130

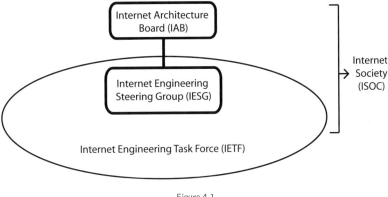

Figure 4.1
ISOC chart

cus the rest of this chapter on those communities and the protocol documents that they produce.

Four groups make up the organizational hierarchy in charge of Internet standardization. They are the Internet Society, the Internet Architecture Board, the Internet Engineering Steering Group, and the Internet Engineering Task Force.[24]

The Internet Society (ISOC), founded in January 1992, is a professional membership society. It is the umbrella organization for the other three groups. Its mission is "to assure the open development, evolution and use of

---

ment Technologies and Techniques of Government: Politics on the Net." Available online at http://duplox.wz-berlin.de/final/jeanette.htm.

24. Another important organization to mention is the Internet Corporation for Assigned Names and Numbers (ICANN). ICANN is a nonprofit organization that has control over the Internet's DNS. Its board of directors has included Vinton Cerf, coinventor of the Internet Protocol and founder of the Internet Society, and author Esther Dyson. "It is ICANN's objective to operate as an open, transparent, and consensus-based body that is broadly representative of the diverse stakeholder communities of the global Internet" (see "ICANN Fact Sheet," available online at http://www.icann.org). Despite this rosy mission statement, ICANN has been the target of intense criticism in recent years. It is for many the central lightning rod for problems around issues of Internet governance. A close look at ICANN is unfortunately outside the scope of this book, but for an excellent examination of the organization, see Milton Mueller's *Ruling the Root* (Cambridge: MIT Press, 2002).

the Internet for the benefit of all people throughout the world."[25] It facilitates the development of Internet protocols and standards. ISOC also provides fiscal and legal independence for the standards-making process, separating this activity from its former U.S. government patronage.

The Internet Architecture Board (IAB), originally called the Internet Activities Board, is a core committee of thirteen, nominated by and consisting of members of the IETF.[26] The IAB reviews IESG appointments, provides oversight of the architecture of network protocols, oversees the standards creation process, hears appeals, oversees the RFC editor, and performs other chores. The IETF (as well as the Internet Research Task Force, which focuses on longer-term research topics) falls under the auspices of the IAB. The IAB is primarily an oversight board, since actually accepted protocols generally originate within the IETF (or in smaller design teams).

Underneath the IAB is the Internet Engineering Steering Group (IESG), a committee of the Internet Society that assists and manages the technical activities of the IETF. All of the directors of the various research areas in the IETF are part of this steering group.

The bedrock of this entire community is the IETF. The IETF is the core area where most protocol initiatives begin. Several thousand people are involved in the IETF, mostly through email lists, but also in face-to-face meetings. "The Internet Engineering Task Force is," in its own words, "a loosely self-organized group of people who make technical and other contributions to the engineering and evolution of the Internet and its technologies."[27] Or elsewhere: "the Internet Engineering Task Force (IETF) is an open global community of network designers, operators, vendors, and researchers producing technical specifications for the evolution of the Internet architecture and the smooth operation of the Internet."[28]

---

25. See http://www.isoc.org.

26. For a detailed description of the IAB, see Brian Carpenter, "Charter of the Internet Architecture Board (IAB)," RFC 2850, BCP 39, May 2000.

27. Gary Malkin, "The Tao of IETF: A Guide for New Attendees of the Internet Engineering Task Force," RFC 1718, FYI 17, October 1993.

28. Paul Hoffman and Scott Bradner, "Defining the IETF," RFC 3233, BCP 58, February 2002.

The IETF is best defined in the following RFCs:

- "The Tao of IETF: A Guide for New Attendees of the Internet Engineering Task Force" (RFC 1718, FYI 17)
- "Defining the IETF" (RFC 3233, BCP 58)
- "IETF Guidelines for Conduct"[29] (RFC 3184, BCP 54)
- "The Internet Standards Process—Revision 3" (RFC 2026, BCP 9)
- "IAB and IESG Selection, Confirmation, and Recall Process: Operation of the Nominating and Recall Committees" (RFC 2727, BCP 10)
- "The Organizations Involved in the IETF Standards Process" (RFC 2028, BCP 11)

These documents describe both how the IETF creates standards and also how the entire community itself is set up and how it behaves.

The IETF is the least bureaucratic of all the organizations mentioned in this chapter. In fact it is not an organization at all, but rather an informal community. It does not have strict bylaws or formal officers. It is not a corporation (nonprofit or otherwise) and thus has no board of directors. It has no binding power as a standards creation body and is not ratified by any treaty or charter. It has no membership, and its meetings are open to anyone. "Membership" in the IETF is simply evaluated through an individual's participation. If you participate via email, or attend meetings, you are a member of the IETF. All participants operate as unaffiliated individuals, not as representatives of other organizations or vendors.

The IETF is divided by topic into various Working Groups. Each Working Group[30] focuses on a particular issue or issues and drafts documents that

---

29. This RFC is an interesting one because of the social relations it endorses within the IETF. Liberal, democratic values are the norm. "Intimidation or ad hominem attack" is to be avoided in IETF debates. Instead IETFers are encouraged to "think globally" and treat their fellow colleagues "with respect as persons." Somewhat ironically, this document also specifies that "English is the de facto language of the IETF." See Susan Harris, "IETF Guidelines for Conduct," RFC 3184, BCP 54, October 2001.

30. For more information on IETF Working Groups, see Scott Bradner, "IETF Working Group Guidelines and Procedures," RFC 2418, BCP 25, September 1998.

are meant to capture the consensus of the group. Like protocols created by other standards bodies, IETF protocols are voluntary standards. There is no technical or legal requirement[31] that anyone actually adopt IETF protocols.

The process of establishing an Internet Standard is gradual, deliberate, and negotiated. Any protocol produced by the IETF goes through a series of stages, called the "standards track." The standards track exposes the document to extensive peer review, allowing it to mature into an RFC memo and eventually an Internet Standard. "The process of creating an Internet Standard is straightforward," they write. "A specification undergoes a period of development and several iterations of review by the Internet community and revision based upon experience, is adopted as a Standard by the appropriate body. . . , and is published."[32]

Preliminary versions of specifications are solicited by the IETF as Internet-Draft documents. Anyone can submit an Internet-Draft. They are not standards in any way and should not be cited as such nor implemented by any vendors. They are works in progress and are subject to review and revision. If they are deemed uninteresting or unnecessary, they simply disappear after their expiration date of six months. They are not RFCs and receive no number.

If an Internet-Draft survives the necessary revisions and is deemed important, it is shown to the IESG and nominated for the standards track. If the IESG agrees (and the IAB approves), then the specification is handed off to the RFC editor and put in the queue for future publication. Cronyism is sometimes a danger at this point, as the old-boys network—the RFC editor, the IESG, and the IAB—have complete control over which Internet-Drafts are escalated and which aren't.

---

31. That said, there are protocols that are given the status level of "required" for certain contexts. For example, the Internet Protocol is a required protocol for anyone wishing to connect to the Internet. Other protocols may be given status levels of "recommended" or "elective" depending on how necessary they are for implementing a specific technology. The "required" status level should not be confused however with mandatory standards. These have legal implications and are enforced by regulatory agencies.

32. Scott Bradner, "The Internet Standards Process—Revision 3," RFC 2026, BCP 9, October 1996.

The actual stages in the standards track are:

1. *Proposed Standard.* The formal entry point for all specifications is here as a Proposed Standard. This is the beginning of the RFC process. The IESG has authority via the RFC editor to elevate an Internet-Draft to this level. While no prior real-world implementation is required of a Proposed Standard, these specifications are generally expected to be fully formulated and implementable.

2. *Draft Standard.* After specifications have been implemented in at least two "independent and interoperable" real-world applications, they can be elevated to the level of a Draft Standard. A specification at the Draft Standard level must be relatively stable and easy to understand. While subtle revisions are normal for Draft Standards, no substantive changes are expected after this level.

3. *Standard.* Robust specifications with wide implementation and a proven track record are elevated to the level of Standard. They are considered to be official Internet Standards and are given a new number in the "STD" subseries of the RFCs (but also retain their RFC number). The total number of Standards is relatively small.

Not all RFCs are standards. Many RFCs are informational, experimental, historic, or even humorous[33] in nature. Furthermore, not all RFCs are full-fledged Standards; they may not be that far along yet.

---

33. Most RFCs published on April 1 are suspect. Take, for example, RFC 1149, "A Standard for the Transmission of IP Datagrams on Avian Carriers" (David Waitzman, April 1990), which describes how to send IP datagrams via carrier pigeon, lauding their "intrinsic collision avoidance system." Thanks to Jonah Brucker-Cohen for first bringing this RFC to my attention. Brucker-Cohen himself has devised a new protocol called "H2O/IP" for the transmission of IP datagrams using modulated streams of water. Consider also "The Infinite Monkey Protocol Suite (IMPS)" described in RFC 2795 (SteQven [sic] Christey, April 2000), which describes "a protocol suite which supports an infinite number of monkeys that sit at an infinite number of typewriters in order to determine when they have either produced the entire works of William Shakespeare or a good television show." Shakespeare would probably appreciate "SONET to Sonnet Translation" (April 1994, RFC 1605), which uses a fourteen-line decasyllabic verse to optimize data transmission over Synchronous Optical Network (SONET). There is also the self-explanatory "Hyper Text Coffee Pot Control Protocol (HTCPCP/1.0)" (Larry Masinter, RFC 2324, April 1998), clearly required reading for any sleep-deprived webmaster.

In addition to the STD subseries for Internet Standards, there are two other RFC subseries that warrant special attention: the Best Current Practice (BCP) documents and informational documents known as FYI.

Each new protocol specification is drafted in accordance with RFC 1111, "Request for Comments on Request for Comments: Instructions to RFC Authors," which specifies guidelines, text formatting and otherwise, for drafting all RFCs. Likewise, FYI 1 (RFC 1150) titled "F.Y.I. on F.Y.I.: Introduction to the F.Y.I. Notes" outlines general formatting issues for the FYI series. Other such memos guide the composition of Internet-Drafts, as well as STDs and other documents. Useful information on drafting Internet standards is also found in RFCs 2223 and 2360.[34]

The standards track allows for a high level of due process. Openness, transparency, and fairness are all virtues of the standards track. Extensive public discussion is par for the course.

Some of the RFCs are extremely important. RFCs 1122 and 1123 outline all the standards that must be followed by any computer that wishes to be connected to the Internet. Representing "the consensus of a large body of technical experience and wisdom,"[35] these two documents outline everything from email and transferring files to the basic protocols like IP that actually move data from one place to another.

Other RFCs go into greater technical detail on a single technology. Released in September 1981, RFC 791 and RFC 793 are the two crucial documents in the creation of the Internet protocol suite TCP/IP as it exists today. In the early 1970s Robert Kahn of DARPA and Vinton Cerf of Stanford Uni-

---

Other examples of ridiculous technical standards include Eryk Salvaggio's "Slowest Modem," which uses the U.S. Postal Service to send data via diskette at a data transfer rate of only 0.0024380952380952380952380952380952380952380 kb/s. He specifies that "[a]ll html links on the diskette must be set up as a href='mailing address' (where 'mailing address' is, in fact, a mailing address)." See Eryk Salvaggio, "Free Art Games #5, 6 and 7," *Rhizome*, September 26, 2000. See also Cory Arcangel's "Total Asshole" file compression system that, in fact, enlarges a file exponentially in size when it is compressed.

34. See Jon Postel and Joyce Reynolds, "Instructions to RFC Authors," RFC 2223, October 1997, and Gregor Scott, "Guide for Internet Standards Writers," RFC 2360, BCP 22, June 1998.

35. Robert Braden, "Requirements for Internet Hosts—Communication Layers," RFC 1122, STD 3, October 1989.

versity teamed up to create a new protocol for the intercommunication of different computer networks. In September 1973 they presented their ideas at the University of Sussex in Brighton and soon afterward finished writing the paper "A Protocol for Packet Network Intercommunication," which was published in 1974 by the IEEE. In that same year Vint Cerf, Yogen Dalal, and Carl Sunshine published "Specification of Internet Transmission Control Program" (RFC 675), which documented details of TCP for the first time. RFC editor Jon Postel and others assisted in the final protocol design.[36] Eventually this new protocol was split in 1978 into a two-part system consisting of TCP and IP. (As mentioned in earlier chapters, TCP is a reliable protocol that is in charge of establishing connections and making sure packets are delivered, while IP is a connectionless protocol that is only interested in moving packets from one place to another.)

One final technology worth mentioning in the context of protocol creation is the World Wide Web. The Web emerged largely from the efforts of one man, the British computer scientist Tim Berners-Lee. During the process of developing the Web, Berners-Lee wrote both HTTP and HTML, which form the core suite of protocols used broadly today by servers and browsers to transmit and display Web pages. He also created the Web address, called a Universal Resource Identifier (URI), of which today's "URL" is a variant: a simple, direct way for locating any resource on the Web.

As Berners-Lee describes it:

The art was to define the few basic, common rules of "protocol" that would allow one computer to talk to another, in such a way that when all computers everywhere did it, the system would thrive, not break down. For the Web, those elements were, in decreasing order of importance, universal resource identifiers (URIs), the Hypertext Transfer Protocol (HTTP), and the Hypertext Markup Language (HTML).[37]

So, like other protocol designers, Berners-Lee's philosophy was to create a standard language for interoperation. By adopting his language, the computers would be able to exchange files. He continues:

---

36. Mueller, *Ruling the Root,* p. 76.

37. Tim Berners-Lee, *Weaving the Web* (New York: HarperCollins, 1999), p. 36.

What was often difficult for people to understand about the design was that there was nothing else beyond URIs, HTTP, and HTML. There was no central computer "controlling" the Web, no single network on which these protocols worked, not even an organization anywhere that "ran" the Web. The Web was not a physical "thing" that existed in a certain "place." It was a "space" in which information could exist.[38]

This is also in line with other protocol scientists' intentions—that an infoscape exists on the Net with no centralized administration or control. (But as I have pointed out, it should not be inferred that a lack of centralized control means a lack of control as such.)

Berners-Lee eventually took his ideas to the IETF and published "Universal Resource Identifiers in WWW" (RFC 1630) in 1994. This memo describes the correct technique for creating and decoding URIs for use on the Web. But, Berners-Lee admitted, "the IETF route didn't seem to be working."[39]

Instead he established a separate standards group in October 1994 called the World Wide Web Consortium (W3C). "I wanted the consortium to run on an open process like the IETF's," Berners-Lee remembers, "but one that was quicker and more efficient. . . . Like the IETF, W3C would develop open technical specifications. Unlike the IETF, W3C would have a small full-time staff to help design and develop the code where necessary. Like industry consortia, W3C would represent the power and authority of millions of developers, researchers, and users. And like its member research institutions, it would leverage the most recent advances in information technology."[40]

The W3C creates the specifications for Web technologies and releases "recommendations" and other technical reports. The design philosophies driving the W3C are similar to those at the IETF and other standards bodies. They promote a distributed (their word is "decentralized") architecture, they promote interoperability in and among different protocols and different end systems, and so on.

In many ways the core protocols of the Internet had their development heyday in the 1980s. But Web protocols are experiencing explosive growth

---

38. Berners-Lee, *Weaving the Web,* p. 36.
39. Berners-Lee, *Weaving the Web,* p. 71.
40. Berners-Lee, *Weaving the Web,* pp. 92, 94.

today. Current growth is due to an evolution of the concept of the Web into what Berners-Lee calls the Semantic Web. In the Semantic Web, information is not simply interconnected on the Internet using links and graphical markup—what he calls "a space in which information could permanently exist and be referred to"[41]—but it is enriched using descriptive protocols that say what the information actually is.

For example, the word "Galloway" is meaningless to a machine. It is just a piece of information that says nothing about what it is or what it means. But wrapped inside a descriptive protocol it can be effectively parsed: "<surname>Galloway</surname>." Now the machine knows that Galloway is a surname. The word has been enriched with semantic value. By making the descriptive protocols more complex, one is able to say more complex things about information, namely, that Galloway is *my* surname, and my given name is Alexander, and so on. The Semantic Web is simply the process of adding extra metalayers on top of information so that it can be parsed according to its semantic value.

Why is this significant? Before this, protocol had very little to do with meaningful information. Protocol does not interface with content, with semantic value. It is, as I have said, against interpretation. But with Berners-Lee comes a new strain of protocol: protocol that cares about meaning. This is what he means by a Semantic Web. It is, as he says, "machine-understandable information."

Does the Semantic Web, then, contradict my earlier principle that protocol is against interpretation? I'm not so sure. Protocols can certainly *say* things about their contents. A checksum does this. A file-size variable does this. But do they actually know the meaning of their contents? So it is a matter of debate as to whether descriptive protocols actually add intelligence to information, or whether they are simply subjective descriptions (originally written by a human) that computers mimic but understand little about. Berners-Lee himself stresses that the Semantic Web is not an artificial intelligence machine.[42] He calls it "well-defined" data, not interpreted data—and

---

41. Berners-Lee, *Weaving the Web*, p. 18.

42. Tim Berners-Lee, "What the Semantic Web Can Represent," available online at http://www.w3.org/DesignIssues/RDFnot.html.

in reality those are two very different things. I promised in the introduction to skip all epistemological questions, and so I leave this one to be debated by others.

As this survey of protocological institutionalization shows, the primary source materials for any protocological analysis of Internet standards are the RFC memos. They began circulation in 1969 with Steve Crocker's RFC "Host Software" and have documented all developments in protocol since.[43] "It was a modest and entirely forgettable memo," Crocker remembers, "but it has significance because it was part of a broad initiative whose impact is still with us today."[44]

While generally opposed to the center-periphery model of communication—what some call the "downstream paradigm"[45]—Internet protocols describe all manner of computer-mediated communication over networks. There are RFCs for transporting messages from one place to another, and others for making sure it gets there in one piece. There are RFCs for email, for webpages, for news wires, and for graphic design.

Some advertise distributed architectures (like IP routing), others hierarchical (like DNS). Yet they all create the conditions for technological innovation based on a goal of standardization and organization. It is a peculiar type of anti-federalism through universalism—strange as it sounds—whereby

---

43. One should not tie Crocker's memo to the beginning of protocol per se. That honor should probably go to Paul Baran's 1964 Rand publication "On Distributed Communications." In many ways it served as the origin text for the RFCs that followed. Although it came before the RFCs and was not connected to it in any way, Baran's memo essentially fulfilled the same function, that is, to outline for Baran's peers a broad technological standard for digital communication over networks.

Other RFC-like documents have also been important in the technical development of networking. The Internet Experiment Notes (IENs), published from 1977 to 1982 and edited by RFC editor Jon Postel, addressed issues connected to the then-fledgling Internet before merging with the RFC series. Vint Cerf also cites the ARPA Satellite System Notes and the PRNET Notes on packet radio (see RFC 2555). There exists also the MIL-STD series maintained by the Department of Defense. Some of the MIL-STDs overlap with Internet Standards covered in the RFC series.

44. Steve Crocker, "30 Years of RFCs," RFC 2555, April 7, 1999.

45. See Minar and Hedlund, "A Network of Peers," p. 10.

universal techniques are levied in such a way as ultimately to revert much decision making back to the local level.

But during this process many local differences are elided in favor of universal consistencies. For example, protocols like HTML were specifically designed to allow for radical deviation in screen resolution, browser type, and so on. And HTML (along with protocol as a whole) acts as a strict standardizing mechanism that homogenizes these deviations under the umbrella of a unilateral standard.

Ironically, then, the Internet protocols that help engender a distributed system of organization are themselves underpinned by adistributed, bureaucratic institutions—be they entities like ICANN or technologies like DNS.

Thus it is an oversight for theorists like Lawrence Lessig (despite his strengths) to suggest that the origin of Internet communication was one of total freedom and lack of control.[46] Instead, it is clear to me that the exact

---

46. In his first book, *Code and Other Laws of Cyberspace* (New York: Basic Books, 1999), Lessig sets up a before/after scenario for cyberspace. The "before" refers to what he calls the "promise of freedom" (p. 6). The "after" is more ominous. Although as yet unfixed, this future is threatened by "an architecture that perfects control" (6). He continues this before/after narrative in *The Future of Ideas: The Fate of the Commons in a Connected World* (New York: Random House, 2001), where he assumes that the network, in its nascent form, was what he calls free—that is, characterized by "an inability to control" (p. 147). Yet "[t]his architecture is now changing" (p. 239), Lessig claims. The world is about to "embrace an architecture of control" (p. 268) put in place by new commercial and legal concerns.

Lessig's discourse is always about a process of becoming, not of always having been. It is certainly correct for him to note that new capitalistic and juridical mandates are sculpting network communications in ugly new ways. But what is lacking in Lessig's work, then, is the recognition that control is endemic to all distributed networks that are governed by protocol. Control was there from day one. It was not imported later by the corporations and courts. In fact distributed networks *must* establish a system of control, which I call protocol, in order to function properly. In this sense, computer networks are and always have been the exact opposite of Lessig's "inability to control."

While Lessig and I clearly come to very different conclusions, I attribute this largely to the fact that we have different objects of study. His are largely issues of governance and commerce while mine are technical and formal issues. My criticism of Lessig is less to deride his contribution, which is inspiring, than to point out our different approaches.

opposite of freedom—that is, control—has been the outcome of the last forty years of developments in networked communications. The founding principle of the Net is control, not freedom. *Control has existed from the beginning.*

Perhaps it is a different type of control than we are used to seeing. It is a type of control based on openness, inclusion, universalism, and flexibility. It is control borne from high degrees of technical organization (protocol), not this or that limitation on individual freedom or decision making (fascism).

Thus it is with complete sincerity that Berners-Lee writes: "I had (and still have) a dream that the web could be less of a television channel and more of an interactive sea of shared knowledge. I imagine it immersing us as a warm, friendly environment made of the things we and our friends have seen, heard, believe or have figured out."[47] The irony is, of course, that in order to achieve this social utopia computer scientists like Berners-Lee had to develop the most highly controlled and extensive mass media yet known. Protocol gives us the ability to build a "warm, friendly" technological space. But it becomes warm and friendly through technical standardization, agreement, organized implementation, broad (sometimes universal) adoption, and directed participation.

I stated in the introduction that protocol is based on a *contradiction* between two opposing machines, one machine that radically distributes control into autonomous locales, and another that focuses control into rigidly defined hierarchies. This chapter illustrates this reality in full detail. The generative contradiction that lies at the very heart of protocol is that *in order to be politically progressive, protocol must be partially reactionary.*

To put it another way, in order for protocol to enable radically distributed communications between autonomous entities, it must employ a strategy of universalization, and of homogeneity. It must be anti-diversity. It must promote standardization in order to enable openness. It must organize peer groups into bureaucracies like the IETF in order to create free technologies.

To be sure, the two partners in this delicate two-step often exist in separate arenas. As protocol pioneer Bob Braden puts it, "There are several vital

---

47. Cited in Jeremie Miller, "Jabber," in *Peer-to-Peer: Harnessing the Power of Disruptive Technologies,* ed. Andy Oram (Sebastapol: O'Reilly, 2001), p. 81.

kinds of heterogeneity."[48] That is to say, one sector can be standardized while another is heterogeneous. The core Internet protocols can be highly controlled while the actual administration of the Net can be highly uncontrolled. Or, DNS can be arranged in a strict hierarchy while users' actual experience of the Net can be highly distributed.

In short, control in distributed networks is not monolithic. It proceeds in multiple, parallel, contradictory, and often unpredictable ways. It is a complex of interrelated currents and counter-currents.

Perhaps I can term the institutional frameworks mentioned in this chapter a type of *tactical standardization,* in which certain short-term goals are necessary in order to realize one's longer-term goals. Standardization is the politically reactionary tactic that enables radical openness. Or to give an example of this analogy in technical terms: DNS, with its hierarchical architecture and bureaucratic governance, is the politically reactionary tactic that enables the truly distributed and open architecture of the Internet Protocol. It is, as Barthes put it, our "Operation Margarine." And this is the generative contradiction that fuels the Net.

---

48. Bob Braden, personal correspondence, December 25, 2002.

III

Protocol Futures

5

Hacking

Disobedience to authority is one of the most natural and healthy acts.
—MICHAEL HARDT AND ANTONIO NEGRI, *Empire*

This book addresses how control exists after decentralization, that is, in specific places where decentralization is done and gone and distribution has set in as the dominant network diagram. Protocol is my answer for that. And as I have shown in the preceding chapters, protocol not only installs control into a terrain that on its surface appears actively to resist it, but in fact goes further to create *the mostly highly controlled mass media hitherto known.*

One reason for its success is the high cost of aberrance levied against those who ignore the global usage of specific technologies. Not to enter into the protocological community carries such a high price that to reject protocol would be foolish. Besides, protocol is an incredibly attractive technology. As protocol pioneer Brian Reid reminisced about his early colleagues: "I felt included by a friendly group of people who recognized that the purpose of networking was to bring everybody in."[1] Protocol is fundamentally a technology of inclusion, and openness is the key to inclusion. "None of the RFCs were ever restricted or classified," wrote Jake Feinler. "This was no mean feat when you consider that they were being funded by [the Department of Defense] during the height of the Cold War."[2]

This fact makes it especially difficult to speak about protocol in a negative sense, for its very success helps preclude outsider positions. Only the participants can connect, and therefore, by definition, there can be *no* resistance to protocol (at least not in any direct or connected sense). *Opposing protocol is like opposing gravity—there is nothing that says it can't be done, but such a pursuit is surely misguided and in the end hasn't hurt gravity much.* While control used to be a law of society, now it is more like a law of nature. Because of this, resisting control has become very challenging indeed.

That said, I move now into the third section of this book, a section called Protocol Futures. In the last section, I called protocol a standardization for tactical purposes only and showed how it is characterized by a political conundrum that involves the acceptance of universal standardization in order to facilitate the ultimate goal of a freer and more democratic medium. In this

---

Epigraph: Michael Hardt and Antonio Negri, *Empire* (Cambridge: Harvard University Press, 2000), p. 210.

1. Cited in Katie Hafner and Matthew Lyon, *Where Wizards Stay Up Late: The Origins of the Internet* (New York: Touchstone, 1996), pp. 144–145.

2. Jake Feinler, "30 Years of RFCs," RFC 2555, April 7, 1999.

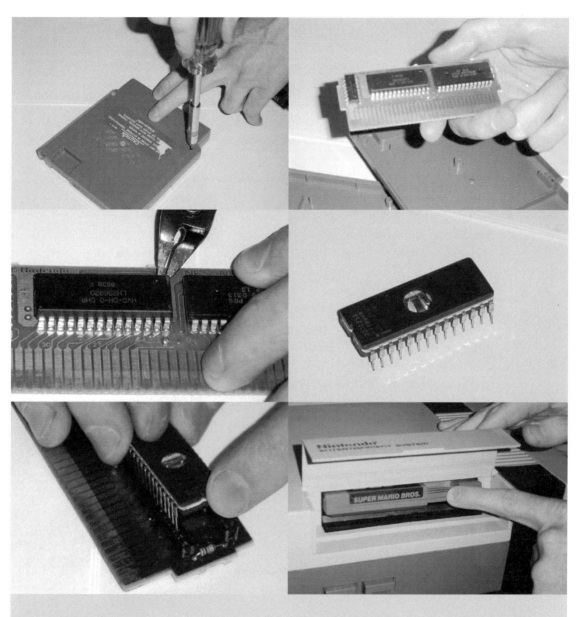

A Super Mario Bros. cartridge for the Nintendo Entertainment System, as hacked by Cory Arcangel. One of the original chips is cut off. Then a new chip is programmed and soldered onto the cartridge.

Cory Arcangel/BEIGE, *Super Mario Clouds* (2002)

section I address a few of the so-called resistive strains within computer culture and how they promise to move protocol into an exciting new space.

But not resistive exactly, for, as I suggest in what follows, the nature of resistance itself has changed within the protocological age. As I discuss in chapter 6 there is a new category of enemy. And this new category of enemy is not at all similar to the bosses, or barons, or bullies of yore. Indeed, part of my inquiry here is to answer the question: what happens when new power structures completely coopt the behavior of their former political adversaries, such that power and the threats to power become indistinguishable? What happens when the enemies of networks are also networks?

As Hardt and Negri write, "this new enemy not only is resistant to the old weapons but actually thrives on them, and thus joins its would-be antagonists in applying them to the fullest."[3] Critical Art Ensemble (CAE) is more pointed in its condemnation of certain (leftist) oppositional techniques that it considers misdirected and (worse) *outdated*. "This situation is particularly ironic," CAE writes, "since the left has always prided itself on using history in critical analysis"—a clear reference to Marxism. "Now rather than acknowledge the present shift in historical forces when constructing strategies for political activism, members of the left continue to act as if they still live in the age of early capital."[4] Political tactics drawn from a bygone age will undoubtedly fail. This is the essence of CAE's argument.[5]

---

3. Hardt and Negri, *Empire,* p. 136.

4. Critical Art Ensemble, *Electronic Civil Disobedience and Other Unpopular Ideas* (New York: Autonomedia, 1996), pp. 9–10. See also "Nomadic Power and Cultural Resistance," in Critical Art Ensemble, *The Electronic Disturbance* (New York: Autonomedia, 1994).

5. While it is important for me to recognize the prescience of CAE, because few others recognized so early and so incisively the nature of politics within the new techno and biopolitical age, several important differences exist between their position and my own. For example, in this passage CAE also conflates decentralized and distributed structures, a distinction I make early and often in this book. While it often uses terms such as "nomadic" that reference distributed systems, CAE claims that "power is decentralized," while I claim here that power is distributed. See Critical Art Ensemble, *Electronic Civil Disobedience and Other Unpopular Ideas,* p. 13. CAE also eschews the lessons from Foucault, Deleuze, and others about the immanent material embodiment of power. It instead claims that "power itself cannot be seen," that "power is experienced only by its effects," and that "[c]apital rarely takes a hard form; like

I suggest then that to live in the age of protocol requires political tactics drawn from within the protocological sphere. "Give up *waiting* for 'the Revolution,'" writes Hakim Bey, but also "give up *wanting* it."[6]

Of course, mine is not an argument against political change, or even against our ability to envision positive action (i.e., utopia). It is rather an indictment of so-called negative political strategies, for in addition to their utter impotence vis-à-vis protocol, it is also clear that protocol is a real step forward in some other sense.

Today there are two things generally said about hackers. They are either terrorists or libertarians. Historically the word meant an amateur tinkerer, an autodidact who might try a dozen solutions to a problem before eking out a success.[7] Aptitude and perseverance have always eclipsed rote knowledge in the hacking community. Hackers are the type of technophiles you like to have around in a pinch, for given enough time they generally can crack any problem (or at least find a suitable kludge).

Thus, as Bruce Sterling writes, the term hacker "can signify the freewheeling intellectual exploration of the highest and deepest potential of computer systems."[8] Or as Steven Levy glowingly reminisces about the original MIT hackers of the early sixties, "they were such fascinating people. . . . Beneath their often unimposing exteriors, they were adventurers, visionaries,

---

power, it exists as an abstraction" (p. 13). A protocological analysis shows that control is almost *never* in abstract form. Rather, protocol ensures that control is literally inscribed onto the very cells and motherboards of bioinformatic networks.

6. Hakim Bey, *T.A.Z.: The Temporary Autonomous Zone, Ontological Anarchy, Poetic Terrorism* (New York: Autonomedia, 1985), p. 101.

7. Robert Graham traces the etymology of the term to the sport of golf: "The word 'hacker' started out in the fourteenth century to mean somebody who was inexperienced or unskilled at a particular activity (such as a golf hacker). In the 1970s, the word 'hacker' was used by computer enthusiasts to refer to themselves. This reflected the way enthusiasts approach computers: they eschew formal education and play around with the computer until they can get it to work. (In much the same way, a golf hacker keeps hacking at the golf ball until they get it in the hole)." See http://www.robertgraham.com/pubs/hacking-dict.html.

8. Bruce Sterling, *The Hacker Crackdown* (New York: Bantam, 1992), p. 51. See also Hugo Cornwall's *Hacker's Handbook* (London: Century, 1988), which characterizes the hacker as a benign explorer. Cornwall's position highlights the differing attitudes between the United States and Europe, where hacking is much less criminalized and in many cases prima facie legal.

risk-takers, artists . . . and the ones who most clearly saw why the computer was a truly revolutionary tool."[9]

These types of hackers are freedom fighters, living by the dictum that data wants to be free.[10] Information should not be owned, and even if it is, non-invasive browsing of such information hurts no one. After all, hackers merely exploit preexisting holes made by clumsily constructed code.[11] And wouldn't the revelation of such holes actually improve data security for everyone involved?

Levy distilled this so-called hacker ethic into several key points:

Access to computers . . . should be unlimited and total.
All information should be free.
Mistrust authority—promote decentralization.
Hackers should be judged by their hacking, not bogus criteria such as degrees, age, race, or position.
You can create art and beauty on a computer.
Computers can change your life for the better.[12]

Several of Levy's points dovetail with my earlier conclusions about protocol. Like the hacker's access to computers, protocol is unlimited and total. Like the hacker's mistrust of authority, protocol also seeks to eliminate arbitrary

---

9. Steven Levy, *Hackers: Heroes of the Computer Revolution* (New York: Anchor Press/Doubleday, 1984), p. ix.

10. This dictum is attributed to Stewart Brand, who wrote that "[o]n the one hand information wants to be expensive, because it's so valuable. The right information in the right place just changes your life. On the other hand, information wants to be free, because the cost of getting it out is getting lower and lower all the time. So you have these two fighting against each other." See *Whole Earth Review,* May 1985, p. 49.

11. Many hackers believe that commercial software products are less carefully crafted and therefore more prone to exploits. Perhaps the most infamous example of such an exploit, one which critiques software's growing commercialization, is the "BackOrifice" software application created by the hacker group Cult of the Dead Cow. A satire of Microsoft's "Back Office" software suite, BackOrifice acts as a Trojan Horse to allow remote access to personal computers running certain versions of Microsoft's Windows operating system.

12. Levy, *Hackers,* pp. 27–33.

authority (yet Levy might have used the word "distribution" rather than "decentralization"). In Levy's view, hackers are respectable, productive members of society. Their goal is merely to understand technology. *"Phrack* is about technology," wrote the editors of *Phrack,* one of the most important hacker magazines, "how to create it, how to use it, and the implications that always arise from it."[13]

Yet after a combination of public technophobia and aggressive government legislation, the identity of the hacker changed in the mid-to-late eighties from do-it-yourself hobbyist to digital outlaw.[14]

The shift hadn't yet come in 1983. That year's *Newsweek* back-to-school issue featured a cover story on Neal Patrick of the 414 hacker group. The article is coy, calling it a "caper" and suggesting that hackers might be more like Robin Hood than outright criminals.[15] Patrick was given immunity by the FBI despite having broken into dozens of computers.

The weather would soon change. A 1985 issue of *Time* described computer viruses as "sabotage."[16] "On March 5, 1986," reported Knight Lightning of *Phrack* magazine, "the following seven phreaks were arrested in what has come to be known as the first computer crime 'sting' operation. Captain Hacker \ Doctor Bob \ Lasertech \ The Adventurer [\] The Highwayman \ The Punisher \ The Warden."[17] "[O]n Tuesday, July 21, 1987," Knight Lightning

---

13. Compaq Disk and Dr. Dude, "Introduction to Diet Phrack (Phrack 36)," *Phrack,* vol. 3, no. 36, file 1, italics not in original.

14. Such legislation includes the Computer Fraud and Abuse Act of 1986, which among other things made it a felony to break into federal computers. For an excellent historical analysis of this transformation, see Sterling's *The Hacker Crackdown.* Andrew Ross explains this transformation by citing, as do Sterling and others, the increase of computer viruses in the late eighties, especially "the viral attack engineered in November 1988 by Cornell University hacker Robert Morris on the national network system Internet. . . . While it caused little in the way of data damage. . . , the ramifications of the Internet virus have helped to generate a moral panic that has all but transformed everyday 'computer culture.'" See Andrew Ross, *Strange Weather: Culture, Science, and Technology in the Age of Limits* (New York: Verso, 1991), p. 75. See also chapter 6 for more on computer viruses.

15. See "Beware: Hackers at Play," *Newsweek,* September 5, 1983.

16. "A Threat from Malicious Software," *Time,* November 4, 1985, p. 94.

17. Knight Lightning, "Shadows Of A Future Past," *Phrack,* vol. 2, no. 21, file 3.

Hypertrophy

The September 11 terrorist attacks brought on a crush of new surveillance activity in the United States. Rumors surfaced that the FBI was installing its notorious email surveillance tool Carnivore willy-nilly on broad civilian networks like Hotmail and AOL. Its expressed purpose was to intercept terror-related communication. As *Wired News* reported on September 12, 2001, "An administrator at one major network service provider said that FBI agents showed up at his workplace on [September 11] 'with a couple of Carnivores, requesting permission to place them in our core.'" Officials at Hotmail were reported to have been "cooperating" with FBI monitoring requests.

On April 6, 2002, the software collective RSG released *Carnivore Personal Edition*, a public domain riff on the original FBI software. The RSG's Carnivore listens to all Internet traffic (email, Web surfing, etc.) on a specific local network. Then, it serves this data stream over the net to an unlimited number of creative interfaces called "clients." The clients are each designed to animate, diagnose, or interpret the network traffic in various ways.

KY [

CAR
ORE CARNIVOI

192.1

205.188

<

! , > 8

2 >v x

VORE CARNIVORE CARNIVO

.1.10:5

192.168.1

RSG, *RSG-CPE0C-1* (2002)

continued, "[a]mong 30–40 others, Bill From RNOC, Eric NYC, Solid State, Oryan QUEST, Mark Gerardo, The Rebel, and Delta-Master have been busted by the United States Secret Service."[18] Many of these hackers (including Knight Lightning himself) were targeted due to their "elite" reputations, a status granted to only the best hackers.

Hackers were deeply discouraged by their newfound identity as outlaws, as exemplified in the famous 1986 hacker manifesto written by someone calling himself[19] The Mentor: "We explore . . . and you call us criminals. We seek after knowledge . . . and you call us criminals."[20] The police crackdown arrived full force in 1990 after the January 15, 1990, crash of AT&T's long-distance system.[21] Because of this transformation, hackers today are commonly referred to as terrorists, ne'er-do-wells who break into computers for personal gain.

The Cuckoo's Egg documents one hacker's exploitation of the honor system existent in open networks, which the author ultimately concludes is utterly unethical, perhaps even criminal: "A hacker's abuse of this openness might mean the end of the casual, communal way the networks are run . . . To have the networks as our playground, we have to preserve our sense of trust; to do that, we have to take it seriously when people break that trust."[22] Even then President William Clinton agreed with this characterization of hackers. In a January 22, 1999, speech, Clinton said: "Revolutions in technology have spread the message and the gifts of freedom but they have also given new opportunities to freedom's enemies. . . . we already are seeing the first wave of deliberate cyber attacks—hackers break into government and business computers, stealing and destroying information, raiding bank accounts, running up credit card charges, extorting money by threats to unleash computer

---

18. Knight Lightning, "The Judas Contract," *Phrack,* vol. 2, no. 22, file 3.

19. While many hackers use gender-neutral pseudonyms, the online magazine *Phrack,* with which The Mentor was associated, was characterized by its distinctly male staff and readership. For a sociological explanation of the gender imbalance within the hacking community, see Paul Taylor, *Hackers: Crime in the Digital Sublime* (New York: Routledge, 1999), pp. 32–42.

20. The Mentor, "The Conscience of a Hacker," *Phrack,* vol. 1, no. 7, file 3, available online at http://www.iit.edu/~beberg/manifesto.html.

21. See Sterling's *The Hacker Crackdown.*

22. Cliff Stoll, *The Cuckoo's Egg* (New York: Pocket Books, 1989), p. 353.

viruses."[23] By the turn of the millennium, the term *hacker* had lost all of its original meaning. Now when people say hacker, they mean terrorist.

As this brief introduction shows, the current debate on hackers is helplessly throttled by the discourse on contemporary liberalism: should one respect data as private property, or should one cultivate individual freedom and leave computer users well enough alone?

Since debates on the relative strengths and weaknesses of liberalism speak very little to the protocological issues discussed in this book (liberalism is a typically modern formation, protocol postmodern; liberalism assumes a rational expressive individual, protocol deals with smooth networks of autonomous entities; and so on), it seems that an analysis of hacking would have little to offer us. "This is perhaps one of the saddest chapters in the history of resistance in the US," writes CAE on the dismal outlook of hacker politics. "Right now the finest political activists are children," it adds, referring to computer hackers who are often teenagers. Then, shifting into a stern parental tone of voice, CAE cautions that "[t]he problem of letting children act as the avant-garde of activism is that they have not yet developed a critical sensibility that would guide them beyond their first political encounter."[24] It seems, then, that hacking has little to tell us about politics in the age of protocological control, and when it does, it tells us that these politics are naive and unproductive.

What I hope to show here is that hacking actually means something else. When viewed allegorically, *hacking is an index of protocological transformations taking place in the broader world of techno-culture.* Hackers do not forecast the death (or avoidance or ignorance) of protocol, but are instead the very harbingers of its assumption.

---

23. William Clinton, cited in "Big Time," *2600* (Spring 1999), p. 4. "The danger from cyber attack has recently received a lot of attention," warned Arizona Senator Jon Kyl in April 2000. "The denial-of-service attacks against popular Internet sites like Yahoo, eBay, and CNN and the arrest earlier this week of a Canadian teenager in the case brought home to Americans just how vulnerable we are." See "Cyber Attack: Improving Prevention and Prosecution," U.S. Congress Subcommittee on Technology, Terrorism, and Government Information, Scottsdale, AZ, April 21, 2000.

24. Critical Art Ensemble, *Electronic Civil Disobedience*, p. 15.

As Hardt and Negri write about the protocological system of control they call Empire: "Empire cannot be resisted by a project aimed at a limited, local autonomy. We cannot move back to any previous social form, nor move forward in isolation. Rather, we must push through Empire to come out the other side."[25] By knowing protocol better than anyone else, hackers push protocol into a state of hypertrophy, hoping to come out the other side. So in a sense, hackers *are created* by protocol, but in another, hackers are protocological actors par excellence.

I turn now to three different moments in the allegory of hacking, three moments that also reveal much about protocol and its future.

Tiger Teams

Sterling writes that the late twentieth century is a moment of transformation from a modern control paradigm based on centralization and hierarchy to a postmodern one based on flexibility and horizontalization:

For years now, economists and management theorists have speculated that the tidal wave of the information revolution would destroy rigid, pyramidal bureaucracies, where everything is top-down and centrally controlled. Highly trained "employees" would take on greater autonomy, being self-starting and self-motivating, moving from place to place, task to task, with great speed and fluidity. "Ad-hocracy" would rule, with groups of people spontaneously knitting together across organizational lines, tackling the problem at hand, applying intense computer-aided expertise to it, and then vanishing whence they came.[26]

From Manuel Castells to Hakim Bey to Tom Peters, this rhetoric has become commonplace. In fact, it is one of the foundational assumptions of this book. Sterling continues by claiming that both hacker groups and the law enforcement officials who track hackers follow this new paradigm: "they *all*

---

25. Hardt and Negri, *Empire,* p. 206. In a spin-off essay derived from this book, I examine the extensive parallels between how empire works in the sociopolitical realm and how protocol works in the technical realm. See Alexander Galloway, "Protocol, or, How Control Exists After Decentralization," *Rethinking Marxism,* vol. 13, no. 3 (Fall 2001).

26. Sterling, *The Hacker Crackdown,* p. 184.

look and act like 'tiger teams' or 'users' groups.' They are all electronic ad-hocracies leaping up spontaneously to attempt to meet a need."[27]

By "tiger teams" Sterling refers to the employee groups assembled by computer companies trying to test the security of their computer systems. Tiger teams, in essence, simulate potential hacker attacks, hoping to find and repair security holes.

The term also evokes the management style known as Toyotism originating in Japanese automotive production facilities. Within Toyotism, small pods of workers mass together to solve a specific problem. The pods are not linear and fixed like the more traditional assembly line, but rather they are flexible and reconfigurable depending on whatever problem might be posed to them.

Management expert Tom Peters notes that the most successful contemporary corporations use these types of tiger teams, eliminating traditional hierarchy within the organizational structure. Documenting the management consulting agency McKinsey & Company, Peters writes: "McKinsey is a huge company. Customers respect it. . . . But there is no traditional hierarchy. There are no organizational charts. No job descriptions. No policy manuals. No rules about managing client engagements. . . . And yet all these things are well understood—make no mistake, McKinsey is not out of control! . . . McKinsey works. It's worked for over half a century."[28] As Sterling suggests, the hacker community also follows this organizational style.

Hackers are autonomous agents that can mass together in small groups to attack specific problems. "ANYONE can write for Phrack Inc.," the influential hacker magazine *Phrack* was keen to point out. "We do not discriminate against anyone for any reason."[29] Flexible and versatile, the hacker pod will

---

27. Sterling, *The Hacker Crackdown,* p. 184.

28. Tom Peters, *Liberation Management: Necessary Disorganization for the Nanosecond Nineties* (New York: Knopf, 1992), pp. 143–144. An older, more decentralized (rather than distributed) style of organizational management is epitomized in Peter Drucker's analysis of General Motors in the thirties and forties. He writes that "General Motors considers decentralization a basic and universally valid concept of order." See Drucker, *The Concept of the Corporation* (New Brunswick: Transaction, 1993), p. 47.

29. "Introduction," *Phrack,* vol. 1, no. 9, phile 1.

often dissolve itself as quickly as it formed and disappear into the network. Thus, what Sterling and others are arguing is that whereby older resistive forces were engaged with "rigid, pyramidal bureaucracies," hackers embody a different organizational management style, one that I have above called protocological.

In this sense, while resistance during the modern age forms around rigid hierarchies and bureaucratic power structures, resistance during the post-modern age forms around the protocological control forces existent in networks. *Hacking means that resistance has changed.*

It is no longer a question of the unified group, but instead a question of the autonomous agent. "Hackers tend to be an idealistic lot. Some might say naive," write the editors of *2600,* a quarterly journal on hacking. "We believe in freedom of speech, the right to explore and learn by doing, and the tremendous power of the *individual.*"[30] Yet this is a new type of individual. This is not the same individual who is the subject of enlightenment liberalism. It is an extension (perhaps) of the modern dream of individuality and independence. Yet this new resistive agent has more in common with the autonomous "boids" of Tom Ray's *Tierra* than with radicals from protocol's prehistory.

"A true hacker is not a group person,"[31] wrote Stewart Brand in 1972. Or, as he would write fifteen years later: "Workers of the world, fan out"[32]—advice that inverts the message of resistance-through-unity found in Marx and Engel's *Communist Manifesto.*

Critical Art Ensemble agrees with this inversion, writing that "[t]he use of power through number—from labor unions to activist organizations—is bankrupt, because such a strategy requires . . . the existence of a centralized present enemy." CAE's "present enemy," I have argued in part I of this book, is in fact distributed rather than centralized. Thus, it makes sense that any forces desiring to resist distributed power should themselves be adept at distributed strategies. I discuss this idea in what follows.

---

30. "The Victor Spoiled," *2600* (Winter 1998–1999), p. 4, emphasis mine.

31. Stewart Brand, "SPACEWAR: Fanatic Life and Symbolic Death Among the Computer Bums," *Rolling Stone,* December 7, 1972, p. 51.

32. Stewart Brand, *The Media Lab: Inventing the Future at MIT* (New York: Viking, 1987), p. 264.

CAE proposes a "nomadic" (rather than sedentary) model for resistance. The nomadic model "seeks to undermine the symbolic order with more ephemeral, process-oriented methods,"[33] it writes. Different nomadic cells, or tiger teams, would coalesce around a specific problem, allowing resistance "to originate from many different points."[34] Then the team would dissolve. Deleuze and Guattari have also written on the nomad as a resistive formation.

This structural form is similar to what Bey refers to in the "temporary autonomous zone" (TAZ). "The TAZ is like an uprising which does not engage directly with the State, a guerilla operation which liberates an area (of land, of time, of imagination) and then dissolves itself to re-form elsewhere/elsewhen, *before* the State can crush it."[35] Hackers exist in such temporary autonomous zones.

The question of groupings is a delicate one because of its close relationship to Marxist politics. "The multitude is biopolitical self-organization,"[36] write Hardt and Negri on their preferred theorization of groupings, the multitude, that is both pluralist and constitutive.

In the hacking community, such pluralistic and constitutive groupings abound. In a 1988 issue of *Phrack* magazine, Knight Lightning listed 130 different hacker groups by name. "Its [sic] literally unbelievable just how many different groups and organizations there are or have been in the phreak/hack/pirate community," he wrote. "The list of 130 groups . . . is probably still just a fraction of the actual amount of groups that there have been"[37] throughout the history of hacking.

Often these tiger teams are labeled conspiracies. But when the anti-hacker crackdown thinks of hacker pods as part of a conspiracy, it is fundamentally misrecognizing the nature of hacker groupings. "[T]he lawmen had gotten used to a word: *conspiracy,*"[38] write Slatalla and Quittner about

---

33. Critical Art Ensemble, *Electronic Civil Disobedience,* p. 38.

34. Critical Art Ensemble, *Electronic Civil Disobedience,* p. 38.

35. Hakim Bey, *T.A.Z.,* p. 101.

36. Hardt and Negri, *Empire,* p. 411.

37. Knight Lightning, "Subdivisions," *Phrack,* vol. 2, no. 23, file 3.

38. Michelle Slatalla and Joshua Quittner, *Masters of Deception: The Gang that Ruled Cyberspace* (New York: HarperCollins, 1995), p. 57.

Access to computers . . . should be unlimited and total.

All information should be free.

Mistrust authority—promote decentralization.

Hackers should be judged by their hacking, not bogus criteria such as degrees, age, race, or position.

You can create art and beauty on a computer.

Computers can change your life for the better.

—Steven Levy, 1984

We explore . . . and you call us criminals. We seek after knowledge . . . and you call us criminals. We exist without skin color, without nationality, without religious bias . . . and you call us criminals . . .

Yes, I am a criminal. My crime is that of curiosity. My crime is that of judging people by what they say and think, not what they look like. My crime is that of outsmarting you, something that you will never forgive me for.

I am a hacker, and this is my manifesto.

—The Mentor, 1986

Revolutions in technology have spread the message and the gifts of freedom but they have also given new opportunities to freedom's enemies . . . we already are seeing the first wave of deliberate cyber attacks—hackers break into government and business computers, stealing and destroying information, raiding bank accounts, running up credit card charges, extorting money by threats to unleash computer viruses.

—Bill Clinton, 1999

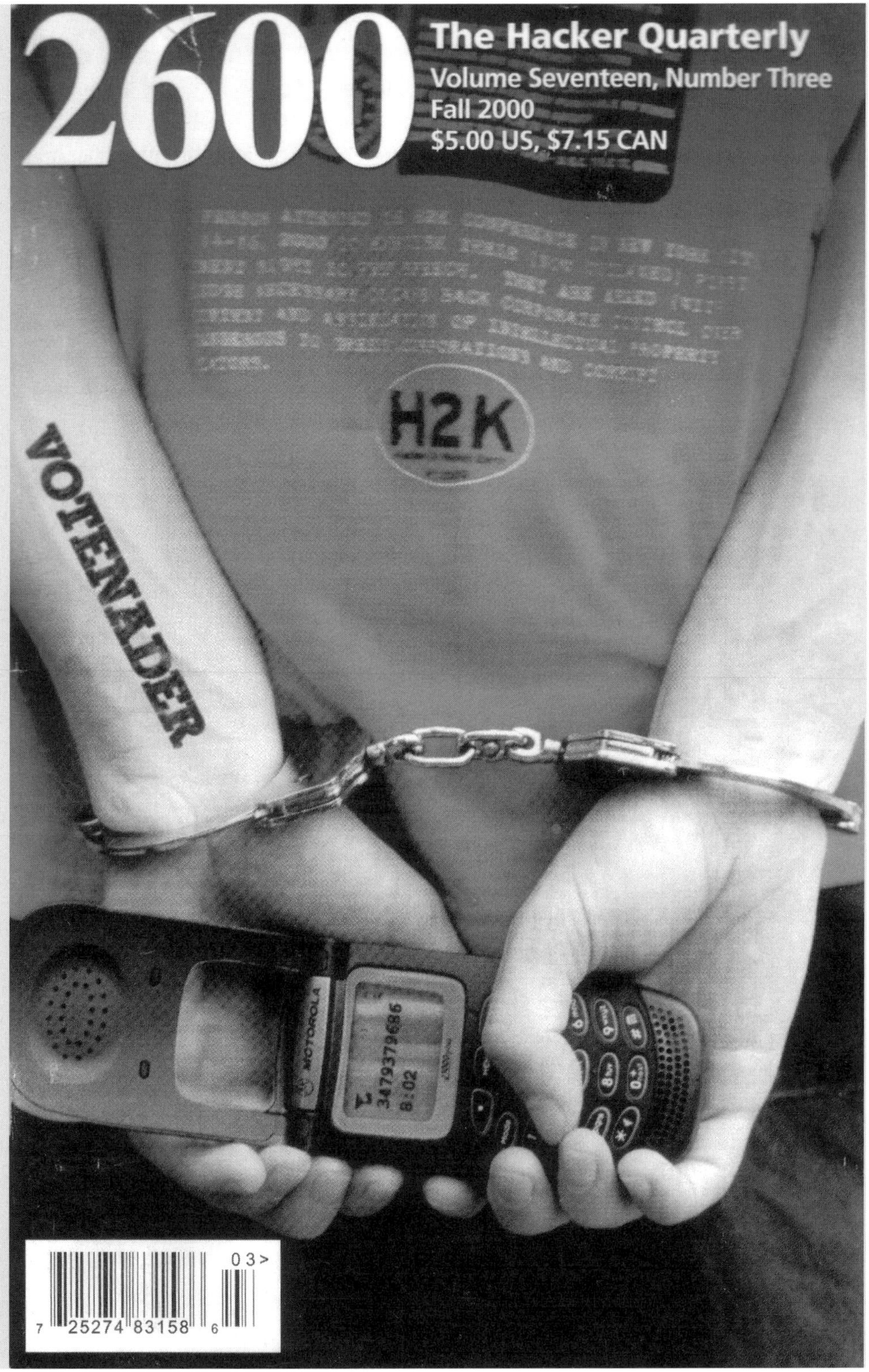
2600
The Hacker Quarterly
Volume Seventeen, Number Three
Fall 2000
$5.00 US, $7.15 CAN
VOTENADER
H2K
MOTOROLA

two anti-hacker investigators whose saga is chronicled in the book *Masters of Deception*. Bruce Sterling writes that many members of hacker teams didn't even know the true names or identities of their cohorts. "This was a highly unconventional profile for a criminal *conspiracy*,"[39] he adds, not without irony. Often there is no center to a hacker tiger team, just a protocological network of autonomous agents.

Code

Hackers know code better than anyone. They speak the language of computers as one does a mother tongue. As I argue in the preface, computer languages and natural languages are very similar. Like the natural languages, computer languages have a sophisticated syntax and grammar. Like the natural languages, computer languages have specific communities and cultures in which the language is spoken. True, computer languages are more often read than spoken. Yet the inclusion of Latin and other so-called dead languages in the pantheon of natural languages proves that languages need not be verbal to be considered natural. Furthermore, computer languages are not mere transcoding schemas like Morse code, or American Sign Language, but have their own semantic specificity as purveyors of meaning.

But if computer languages are so similar to natural languages, why are they virtually ignored by linguists and literary critics? What makes code so much different from normal language?

In 1967 the artist Sol LeWitt hit upon a possible answer when he defined the process of conceptual art: "In conceptual art the idea or concept is the most important aspect of the work. When an artist uses a conceptual form of art, it means that all of the planning and decisions are made beforehand and the execution is a perfunctory affair. The idea becomes a machine that makes the art."[40] LeWitt's perspective on conceptual art has important implications for code, for in his estimation conceptual art is nothing but a type of

---

39. Sterling, *The Hacker Crackdown,* pp. 91–92, emphasis mine.

40. Sol LeWitt, "Paragraphs on Conceptual Art," in *Conceptual Art: A Critical Anthology* ed. Alexander Alberro and Blake Stimson, (Cambridge: MIT Press, 1999), p. 12. Thanks to Mark Tribe for bring this passage to my attention.

code for art making. LeWitt's art is an algorithmic process. The algorithm is prepared in advance, then later executed by the artist (or another artist, for that matter).

Code thus purports to be multidimensional. Code draws a line between what is material and what is active, in essence saying that writing (hardware) cannot *do* anything, but must be transformed into code (software) to be effective.

Northrop Frye says a very similar thing about language when he writes that the process of literary critique essentially creates a metatext, outside the original source material, that contains the critic's interpretations of that text.[41] In fact Kittler defines software itself as precisely that "logical abstraction" that exists in the negative space between people and the hardware they use.[42] Katherine Hayles has also reflected on the multidimensionality of digital signs. Her term "flickering signifiers" shows that digital images are the visible manifestations of underlayers of code often hidden.

But how can code be so different from mere writing? The answer to this lies in the unique nature of computer code. It lies not in the fact that code is sublinguistic, but rather in the fact that it is *hyper*linguistic. Code is a language, but a very special kind of language. *Code is the only language that is executable.*

As Kittler has pointed out, "There exists no word in any ordinary language which does what it says. No description of a machine sets the machine into motion."[43] The imperative voice (and what philosophers like Austin, Searle, and others talk about in the area of speech act theory) attempts to affect change through persuasion but has little real material affect.[44] So code

---

41. See Northrop Frye, *Anatomy of Criticism* (Princeton: Princeton University Press, 1957). See also Fredric Jameson's engagement with this same subject in "From Metaphor to Allegory," in *Anything,* ed. Cynthia Davidson (Cambridge: MIT Press, 2001).

42. Friedrich Kittler, "On the Implementation of Knowledge—Toward a Theory of Hardware," *Nettime,* February 6, 1999.

43. Kittler, "On the Implementation of Knowledge."

44. Saying this will doubtless get me into hot water with the philosophers. But Kittler's point is well taken: If you yell fire in a crowded theater it may come to pass that people flee, but there is no material logic that necessitates they will. On the other had, code *is* based on material logic, so it is predisposed to affect material change.

is the first language that actually does what it says—it is a machine for converting meaning into action.[45]

Code has a semantic meaning, but it also has an enactment of meaning. Thus, while natural languages such as English or Latin only have a legible state, code has both a legible state and an executable state. In this way, code is the summation of language plus an executable metalayer that encapsulates that language.

Sterling reflects on the conglomeration of these different layers of code into discrete pieces of software:

The stuff we call "software" is not like anything that human society is used to thinking about. Software is something like a machine, and something like mathematics, and something like language, and something like thought, and art, and information . . . but software is not in fact any of those other things. The protean quality of software is one of the greatest sources of its fascination. It also makes software very powerful, very subtle, very unpredictable, and very risky.[46]

In the same way that natural languages have enjoyed a hierarchically privileged relationship over computer languages, it is common to group different types of computer languages in a hierarchy, as Kittler does in his essay "There Is No Software":

Programming languages have eroded the monopoly of ordinary language and grown into a new hierarchy of their own. This postmodern Babylonian tower reaches from simple operation codes whose linguistic extension is still a hardware configuration passing through an assembler whose extension is this very opcode up to high-level programming languages whose extension is that very assembler.[47]

---

45. For an interesting commentary on the aesthetic dimensions of this fact, see Geoff Cox, Alex McLean, and Adrian Ward's "The Aesthetics of Generative Code," available online at http://sidestream.org/papers/aesthetics.

46. Sterling, *The Hacker Crackdown,* p. 31.

47. Friedrich Kittler, "There Is No Software," in *Electronic Culture,* ed. Timothy Druckrey (New York: Aperture, 1996), p. 332.

Because machine code is a direct tracing of the actual machinic circuitry existent in one's hardware, it is perceived to be the most fundamental type of computer language. By contrast, because high-level languages like C++ or Perl must be compiled/interpreted, they are considered to be less fundamental.

Yet such a division is perhaps misguided due to the logical nature of computer software. Compilers, linkers, interpreters, and the like are fundamentally translation machines, machines that move code from one form to another. Uncompiled source code is *logically* equivalent to that same code compiled into assembly language and/or linked into machine code. For example, it is absurd to claim that a certain value expressed as a hexadecimal (base 16) number is more or less fundamental than that same value expressed as a binary (base 2) number. They are simply two expressions of the same value. By the same token, it is wrong to make claims about the relative purity of machine code over source code, for each is equally mediated by the environment of the digital computer. Source code might be simpler for humans to read, but this says very little about its relationship in a larger hierarchy of code, of which there can be none.[48]

The hacker's close relationship to code displays the power of protocol, particularly its ability to compel autonomous actors toward a more vital or affective state within their particular distributed milieu.

Possibility

Protocol is synonymous with possibility. This was demonstrated in part I of this book. Protocol outlines the playing field for what can happen, and where. If one chooses to ignore a certain protocol, then it becomes impossible to communicate on that particular channel. No protocol, no connection.

Any newbie hacker will be able to tell you that hacking relies on "exploits," preexisting bugs that are leveraged by the hacker to gain access to a computer. Burglars know that houses have their own exploits. Locks may be

---

48. One difficulty in this argument is that with certain computer languages (such as C) it is much more difficult to convert an executable file back into source code than it is to convert source code into an executable file. This limitation has been leveraged by commercial software companies, whose business is based on the proprietary control of source code (Microsoft, etc.).

picked, windows broken, doors jimmied. The same can be done to a computer: buffers may be overflowed, trapdoors sprung, Trojan Horses deployed. Yet while a burglar's exploits often rely on physical force to gain entry, a hacker's exploits generally rely on *logical* force. That is, while physical hacks are often necessary, hackers generally focus on the bugs and loopholes of a machine's logical code base.[49]

Protocol *is synonymous with possibility.* From the perspective of protocol, if you can do it, it can't be bad, because if it were bad, then it would have been outlawed years ago by protocol.

Hackers don't care about rules, feelings, or opinions. They care about what is true and what is possible. And in the logical world of computers, if it is possible then it is real. *Can* you break into a computer, not *should* you or is it *right* to. When poured in a vessel, water will fill the vessel completely; when poured into a computer network, the hacker will enter any space available to him.

In fact, possibility often erases the unethical in the mind of the hacker. An anecdote from the legendary hacker Acid Phreak illustrates this well. After being told certain personal details about his rhetorical opponent John Perry Barlow, information that he would later use to obtain Barlow's credit history, Acid Phreak screamed, "Mr. Barlow: Thank you for posting all I need to know to get your credit information and a whole lot more! Now, who is to blame? ME for getting it or YOU for being such an idiot?!"[50] Most hackers would answer: You, for being such an idiot.

Jameson said somewhere that one of the most difficult things to do under contemporary capitalism is to envision utopia. This is precisely why possibility is important. Deciding (and often struggling for) what is possible is

---

49. A primary counterexample is the practice of social engineering, whereby the hacker leaves the logical realm and enters the physical world to glean passwords and other important information. There are a few very rare examples of specific hacks that destroy hardware through abuse of mechanical parts such as the disk drive. The Electrohome color vector monitor used in Sega arcade games like Tac/Scan can actually catch on fire if misused (thanks to Mark Daggett for bringing this to my attention). I also ran across an urban myth describing an obsolete model of computer that would light on fire as the result of a specific key combination, but this could not be confirmed.

50. Slatalla and Quittner, *Masters of Deception,* p. 102.

the first step in a utopian vision based in desire, based in what one *wants*. Hackers are machines for the identification of this possibility.

Pierre Lévy is one writer who has been able to articulate eloquently the possibility of utopia in the cyberspace of digital computers.[51] "Cyberspace," he writes, "brings with it methods of perception, feeling, remembering, working, of playing and being together. . . . The development of cyberspace . . . is one of the principal aesthetic and political challenges of the coming century."[52] Lévy's visionary tone is exactly what Jameson warns is lacking in much contemporary discourse.

The relationship between utopia and possibility is a close one. It is necessary to know what one wants, to know what is *possible* to want, before a true utopia may be envisioned. "When computers become available to everybody," wrote Stewart Brand in 1972, "the hackers take over: We are all Computer Bums, all more empowered as individuals and as cooperators."[53] Or as McKenzie Wark writes, "Whatever code we hack, be it programming language, poetic language, math or music, curves or colourings, we create the possibility of new things entering the world."[54] Thus, I suggest that the hacker's unique connection to the realm of the possible, via protocol that structures itself on precisely that threshold of possibility, gives the hacker special insight into the nature of utopia—what he or she *wants* out of computers.

One of the most important signs of this utopian instinct is the hacking community's anti-commercial bent. Software products have long been developed and released into the public domain, with seemingly no profit motive on the side of the authors, simply for the higher glory of the code itself.

"Spacewar was not sold," Steven Levy writes, referring to the early video game developed by several early computer enthusiasts at MIT. "Like any other

---

51. Another is the delightfully schizophrenic Ted Nelson, inventor of hypertext. See *Computer Lib/Dream Machines* (Redmond, WA: Tempus/Microsoft, 1987).

52. Pierre Lévy, *L'intelligence collective: Pour une anthropologie du cyberspace* (Paris: Éditions la Découverte, 1994), p. 120.

53. Stewart Brand, "SPACEWAR: Fanatic Life and Symbolic Death among the Computer Bums," *Rolling Stone,* December 7, 1972, p. 58.

54. McKenzie Wark, "A Hacker Manifesto [version 4.0]," available online at http://subsol.c3.hu/subsol_2/contributors0/warktext.html.

program, it was placed in the drawer for anyone to access, look at, and rewrite as they saw fit."[55]

The limits of personal behavior become the limits of possibility to the hacker. Thus, it is obvious to the hacker that one's personal investment in a specific piece of code can do nothing but hinder that code's overall development.

"Sharing of software . . . is as old as computers," writes free software guru Richard Stallman, "just as sharing of recipes is as old as cooking."[56]

Code does not reach its apotheosis *for people,* but exists within its own dimension of perfection. The hacker feels obligated to remove all impediments, all inefficiencies that might stunt this quasi-aesthetic growth. "In its basic assembly structure," writes Andrew Ross, "information technology involves processing, copying, replication, and simulation, and therefore does not recognize the concept of private information property."[57] Commercial ownership of software is the primary impediment hated by all hackers because it means that code is limited—limited by intellectual property laws, limited by the profit motive, limited by corporate "lamers."

Even Kevin Mitnick, a hacker maligned by some for his often unsavory motivations, admits that the code itself has a higher priority than any commercial motivation:

You get a better understanding of the cyberspace, the computer systems, the operating systems, how the computer systems interact with one another, that basically, was my motivation behind my hacking activity in the past, it was just from the gain of knowledge and the thrill of adventure, nothing that was well and truly sinister such as trying to get any type of monetary gain or anything.[58]

---

55. Levy, *Hackers,* p. 53. In his 1972 *Rolling Stone* article on the game, Stewart Brand went so far as to publish Alan Kay's source code for Spacewar right alongside his own article, a practice rarely seen in popular publications. See Brand, "SPACEWAR," p. 58.

56. Richard Stallman, "The GNU Project," available online at http://www.gnu.org/gnu/thegnuproject.html and in Chris Dibona et al, eds., *Open Sources: Voices from the Open Source Revolution* (Sebastopol, CA: O'Reilly, 1999).

57. Ross, *Strange Weather,* p. 80.

58. From a telephone interview with Kevin Mitnick, cited in Taylor, *Hackers,* p. 57. For more details on the Mitnick story, see the following texts: Katie Hafner and John Markoff, *Cyberpunk: Outlaws and Hackers on the Computer Frontier* (New York: Touchstone, 1991); Tsu-

A British hacker named Dr-K hardens this sentiment into an explicit anti-commercialism when he writes that "[c]orporations and government cannot be trusted to use computer technology for the benefit of ordinary people."[59]

It is for this reason that the Free Software Foundation was established in 1985. It is for this reason that so much of the non-PC computer community is dominated by free, or otherwise de-commercialized software.[60] The hacker ethic thus begets utopia simply through its rejection of all commercial mandates.

However, greater than this anti-commercialism is a pro-protocolism. Protocol, by definition, is *open source,* the term given to a technology that makes public the source code used in its creation. That is to say, protocol is nothing but an elaborate instruction list of how a given technology should work, from the inside out, from the top to the bottom, as exemplified in the RFCs described in chapter 4.

While many closed source technologies appear to be protocological due to their often monopolistic position in the market place, a true protocol cannot be closed or proprietary. It must be paraded into full view before all, and agreed to by all. It benefits over time through its own technological development in the public sphere. It must exist as pure, transparent code (or a pure *description* of how to fashion code).

As concerned protocological actors, hackers have often called attention to commercial or governmental actions that impede protocol through making

<hr />

tomu Shimomura, *Takedown: The Pursuit and Capture of Kevin Mitnick, America's Most Wanted Computer Outlaw—By the Man Who Did It* (New York: Hyperion, 1996); Jonathan Littman, *The Fugitive Game: Online With Kevin Mitnick* (New York: Little Brown & Co, 1997); Jeff Goodell, *The Cyberthief and the Samurai: The True Story of Kevin Mitnick—And the Man Who Hunted Him Down* (New York: Dell, 1996). See also the documentary *Freedom Downtime* available from *2600* magazine. Mitnick himself has written a book on security called *The Art of Deception* (New York: John Wiley & Sons, 2002).

59. Dr-K, *A Complete H@cker's Handbook: Everything You Need to Know About Hacking in the Age of the Internet* (London: Carlton, 2000), p. 9.

60. The primary example of this trend is the free, open source operating system Linux. Virtually every software product has a freeware analog that often performs better than its commercial counterpart. Examples include Apache, a free Web server; MySQL, a free relational database that competes with high-end commercial databases such as Oracle; and Perl, a free open source scripting language.

certain technologies proprietary or opaque. One such impediment is the Digital Millennium Copyright Act (DMCA) of 1998. The hacker journal *2600* has pointed out that the DMCA "basically makes it illegal to reverse engineer technology," reverse engineering being the term that describes the interpellation of source code through an examination of the results of that code. "This means that you're not allowed to take things apart and figure out how they work if the corporate entities involved don't want you to."[61] This certainly is a pity for those wishing free use of commercial technology products, however it is a greater pity for protocol. For if technology is proprietary it ceases to be protocological.

The synonym for possibility most commonly used in today's technospeak is *access*. On the Net, something is possible only if it is accessible. Hackers reject situations where access to technology is limited. Purveyors of proprietary technology "want to be able to dictate how, when, and where you can access content,"[62] complain the editors of *2600* over a lawsuit levied by the Motion Picture Association of America against hackers who had cracked the proprietary limitations of the DVD media format.

2600 writes, correctly, that the real issue here is one of control over a specific technical knowledge, not potential piracy of DVD media: "The Motion Picture Association of America wanted to make sure they had *control* and that nobody, not hackers, not civil libertarians, not ordinary people in the street—dared to figure out how to challenge that control. Selling a pirated movie is nothing to them. But telling people how the technology works is the real threat."[63]

What hacking reveals, then, is not that systems are secure or insecure, or that data wants to be free or proprietary, but that with protocol comes the exciting new ability to leverage possibility and action through code. That it does not show this on its surface comes as no surprise.

It has been the goal then of this chapter to allegorize hacking such that what was once a very boring discourse on ethics and liberalism may be utilized in the greater development of a theory of protocol.

---

61. "The Next Chapter," *2600* (Spring 2000), p. 6.

62. "A Summer of Trials," *2600* (Fall 2000), p. 4.

63. "Direction," *2600* (Winter 2000–2001), p. 4.

6

Tactical Media

The Internet is like the *Titanic.* It is an instrument which performs
extraordinarily well but which contains its own catastrophe.
—PAUL VIRILIO, "Infowar"

Like many interesting social movements that may manifest themselves in a variety of ways, tactical media has an orthodox definition and a more general one. The orthodoxy comes from the new tech-savvy social movements taking place in and around the Western world and associated with media luminaries such as Geert Lovink, Ricardo Dominguez (with the Electronic Disturbance Theater), and Critical Art Ensemble. Tactical media is the term given to political uses of both new and old technologies, such as the organization of virtual sit-ins, campaigns for more democratic access to the Internet, or even the creation of new software products not aimed at the commercial market.

"Tactical Media are what happens when the cheap 'do it yourself' media, made possible by the revolution in consumer electronics and expanded forms of distribution (from public access cable to the internet) are exploited by groups and individuals who feel aggrieved by or excluded from the wider culture," write tactical media gurus David Garcia and Geert Lovink. "Tactical media are media of crisis, criticism and opposition."[1] Thus, tactical media means the bottom-up struggle of the networks against the power centers. (And of course the networks against the power centers who have recently reinvented themselves as networks!)

But there is also a more general way of thinking about tactical phenomena within the media. That is to say, there are certain *tactical effects* that often leave only traces of their successes to be discovered later by the ecologists of the media. This might include more than would normally fit under the orthodox definition. Case in point: computer viruses. In a very bland sense, they are politically bankrupt and certainly no friend of the tactical media practitioner. But in a more general sense they speak volumes on the nature of network-based conflict.

For example computer viruses are incredibly effective at identifying anti-protocological technologies. They infect proprietary systems and propagate

---

Epigraphs: Paul Virilio, "Infowar," in *Ars Electronica,* ed. Timothy Druckrey (Cambridge: MIT Press, 1999), p. 334. One assumes that the italicized *"Titanic"* may refer to James Cameron's 1997 film as well as the fated passenger ship, thereby offering an interesting double meaning that suggests, as others have aptly argued, that films, understood as texts, contain their own undoing. John Arquilla and David Ronfeldt, *Networks and Netwars: The Future of Terror, Crime, and Militancy* (Santa Monica, CA: Rand, 2001), p. 6.

1. David Garcia and Geert Lovink, "The ABC of Tactical Media," *Nettime,* May 16, 1997.

through the homogeneity contained within them. Show me a computer virus and I'll show you proprietary software with a market monopoly.

I will not repeat here the excellent attention given to the subject by CAE, Lovink, and others. Instead in this chapter I would like to examine tactical media as *those phenomena that are able to exploit flaws in protocological and proprietary command and control, not to destroy technology, but to sculpt protocol and make it better suited to people's real desires.* "Resistances are no longer marginal, but active in the center of a society that opens up in networks,"[2] Hardt and Negri remind us. Likewise, techno-resistance is not outside protocol but at its center. Tactical media propel protocol into a state of hypertrophy, pushing it further, in better and more interesting ways.

Computer Viruses

While a few articles on viruses and worms appeared in the 1970s and beginning of the 1980s,[3] Frederick Cohen's work in the early 1980s is cited as the first sustained examination of computer viruses. He approached this topic from a scientific viewpoint, measuring infection rates, classifying different types of viruses, and so on.

The record for the smallest virus is a Unix "sh" command script. In the command interpreter of Unix, you can write a virus that takes only about 8 characters. So, once you are logged into a Unix system, you can type a 8 character command, and before too long, the virus will spread. That's quite small, but it turns out that with 8 characters, the virus can't do anything but reproduce. To get a virus that does interesting damage, you need around 25 or 30 characters. If you want a virus that evolves, replicates, and does damage, you need about 4 or 5 lines.[4]

---

2. Michael Hardt and Antonio Negri, *Empire* (Cambridge: Harvard University Press, 2000), p. 25.

3. Ralf Burger cites two articles: "ACM Use of Virus Functions to Provide a Virtual APL Interpreter Under User Control" (1974), and John Shoch and Jon Huppas's "The Worm Programs—Early Experience with a Distributed Computation" (1982), which was first circulated in 1980 in abstract form as "Notes on the 'Worm' programs" (IEN 159, May 1980). See Ralf Burger, *Computer Viruses* (Grand Rapids: Abacus, 1988), p. 19.

4. Frederick Cohen, *A Short Course on Computer Viruses* (New York: John Wiley & Sons, 1994), p. 38.

Cohen first presented his ideas on computer viruses to a seminar in 1983. His paper "Computer Viruses—Theory and Experiments" was published in 1984, and his Ph.D. dissertation titled "Computer Viruses" (University of Southern California) in 1986.

Cohen defines a computer virus as "a program that can 'infect' other programs by modifying them to include a, possibly evolved, version of itself."[5] Other experts agree: "a virus is a self-replicating code segment which must be attached to a host executable."[6] Variants in the field of malicious code include worms and Trojan horses. A worm, like a virus, is a self-replicating program but one that requires no host to propagate. A Trojan horse is a program that appears to be doing something useful but also executes some piece of undesirable code hidden to the user.

In the literature viruses are almost exclusively characterized as hostile or harmful. They are often referred to completely in the negative, as in "antivirus software" or virus prevention, or as one author calls it, a "high-tech disease." They are considered nearly exclusively in the context of detection, interception, identification, and removal.

Why is this the case? Viral marketing, emergent behavior, self-replicating systems—these concepts are all the rage at the turn of the millennium. Yet computer viruses gain from none of these positive associations. They are thought of as a plague used by terrorists to wreak havoc on the network.

So why did computer viruses become so closely connected with the viral metaphor in biology? Why think of self-replicating programs as a "virus" and not simply a parasitic nuisance, or a proper life form? Even the father of computer virus science, Cohen, thought of them as a form of artificial life[7]

---

5. Cohen, *A Short Course on Computer Viruses*, p. 2.

6. W. Timothy Polk et al., *Anti-Virus Tools and Techniques for Computer Systems* (Park Ridge, NJ: Noyes Data Corporation, 1995), p. 4.

7. Indeed pioneering viral scientist Frederick Cohen is the most notable exception to this rule. He recognized the existence of "benevolent viruses" that perform maintenance, facilitate networked applications, or simply live in "peaceful coexistence" with humans: "I personally believe that reproducing programs are living beings in the information environment." See Cohen, *A Short Course on Computer Viruses*, pp. 159–160, 15–21, and Frederick Cohen, *It's Alive!* (New York: John Wiley & Sons, 1994). The author Ralf Burger is also not completely

and recognized the limitations of the biological analogy. "[C]onsider a biological disease that is 100% infectious, spreads whenever animals communicate, kills all infected animals instantly at a given moment, and has no detectable side effect until that moment,"[8] wrote Cohen, identifying the ultimate inaccuracy of the analogy. How did self-replicating programs *become viruses?*

For example, if viruses had emerged a decade later, in the late 1990s, it is likely that they would have a completely difference sociocultural meaning. They would most certainly be thought of more as a distributed computing system (like SETI@home) or an artificial life experiment (like Tom Ray's *Tierra*), or an artwork (like Mark Daggett's email worm, *vcards*), or as a nuisance (spam), or as a potential guerilla marketing tool (adware)—not a biological infestation.

Computer viruses acquired their current discursive position because of a unique transformation that transpired in the mid-1980s around the perception of technology. In fact several phenomena, including computer hacking, acquired a distinctly negative characterization during this period of history because of the intense struggle waging behind the scenes between proprietary and protocological camps.

My hypothesis is this: Early on, computer viruses were identified with the AIDS epidemic. It is explicitly referenced in much of the literature on viruses, making AIDS both the primary biological metaphor and primary social anxiety informing the early discourse on computer viruses. In that early mode, the virus itself was the epidemic. Later, the discourse on viruses turned toward weaponization and hence terrorism. Here, the virus author is the epidemic. Today the moral evaluation of viruses is generally eclipsed by the search for their *authors,* who are prosecuted as criminals and often terrorists. The broad viral epidemic itself is less important than the *criminal mind*

---

pessimistic, instructing readers that when "used properly, [viruses] may bring about a new generation of self-modifying computer operating systems. . . . Those who wish to examine and experiment with computer viruses on an experimental level will quickly discover what fantastic programming possibilities they offer." See Burger, *Computer Viruses,* p. 2.

8. Frederick Cohen, "Implications of Computer Viruses and Current Methods of Defense," in *Computers Under Attack: Intruders, Worms, and Viruses,* ed. Peter Denning (New York: ACM, 1990), p. 383.

that brings it into existence (or the flaws in proprietary software that allow it to exist in the first place).

Thus, by the late 1990s viruses are the visible indices of a search for evildoers within technology, not the immaterial, anxious fear they evoked a decade earlier with the AIDS crisis.

Computer viruses appeared in a moment in history where the integrity and security of bodies, both human and technological, was considered extremely important. Social anxieties surrounding both AIDS and the war on drugs testify to this. The AIDS epidemic in particular is referenced in much of the literature on viruses.[9] This makes sense because of the broad social crisis created by AIDS in the mid-to-late 1980s (and beyond). "In part," writes Ralf Burger, "it seems as though a hysteria is spreading among computer users which nearly equals the uncertainty over the AIDS epidemic."[10] A good example of this discursive pairing of AIDS and computer viruses is seen in the February 1, 1988, issue of *Newsweek*. Here an article titled "Is Your Computer Infected?," which reports on computer viruses affecting hospitals and other institutions, is paired side-by-side with a medical article on AIDS.

Consider two examples of this evolving threat paradigm. The Jerusalem virus[11] was first uncovered in December 1987 at the Hebrew University of Jerusalem in Israel. "It was soon found that the virus was extremely widespread, mainly in Jerusalem, but also in other parts of the country, especially in the Haifa area,"[12] wrote professor Yisrael Radai. Two students, Yuval Rakavy and Omri Mann, wrote a counterprogram to seek out and delete the virus.

Mystery surrounds the origins of the virus. As Cohen writes, terrorists are suspected of authoring this virus. It was timed to destroy data precisely on

---

9. See Philip Fites, Peter Johnson, and Martin Kratz, *The Computer Virus Crisis* (New York: Van Nostrand Reinhold, 1992), pp. 28, 54, 105–117, 161–162; Burger, *Computer Viruses,* p. 1; Charles Cresson Wood, "The Human Immune System as an Information Systems Security Reference Model," in *Rogue Programs,* ed. Lance Hoffman (New York: Van Nostrand Reinhold, 1990), pp. 56–57. In addition, the AIDS Info Disk, a Trojan horse, is covered in almost every book on the history of computer viruses.

10. Burger, *Computer Viruses,* p. 1.

11. Also called the "Israeli" or "PLO" virus.

12. Yisrael Radai, "The Israeli PC Virus," *Computers and Security* 8, no. 2, (1989), p. 112.

the first Friday the thirteenth it encountered, which landed on May 13, 1988, and coincided with the day commemorating forty years since the existence of a Palestinian state.[13] (A subsequent outbreak also happened on Friday, January 13, 1989 in Britain.) The *Edmonton Journal* called it the work of a "saboteur." This same opinion was voiced by *The New York Times,* who reported that the Jerusalem virus "was apparently intended as a weapon of political protest."[14] Yet Radai claims that in subsequent, off-the-record correspondence, the *Times* reporter admitted that he was "too quick to assume too much about this virus, its author, and its intent."[15]

In the end it is of little consequence whether or not the virus was written by the Palestine Liberation Organization (PLO). What matters is that this unique viral threat was menacing enough to influence the judgment of the media (and also Cohen) to believe, and perpetuate the belief, that viruses have a unique relationship to terrorists. Words like "nightmare," "destroy," "terrorist," and "havoc" pervade the *Times* report.

Second, consider the "AIDS Information Introductory Diskette Version 2.0" Disk. On December 11, 1989, the PC Cyborg Corporation mailed approximately 10,000[16] computer diskettes to two direct mail lists compiled from the subscribers to *PC Business World* and names from the World Health Organization's 1988 conference on AIDS held in Stockholm.[17] The disk, which carried the title "AIDS Information Introductory Diskette Version 2.0," presented an informational questionnaire to the user and offered an assessment of the user's risk levels for AIDS based on his or her reported behavior.

The disk also acted as a Trojan horse containing a virus. The virus damages file names on the computer and fills the disk to capacity. The motives of the virus author are uncertain in this case, although it is thought to be a

---

13. Cohen, *A Short Course on Computer Viruses,* p. 45.

14. "Computer Systems Under Siege, Here and Abroad," *The New York Times,* January 31, 1988, section 3, p. 8.

15. Cited in Radai, "The Israeli PC Virus," p. 113.

16. Frederick Cohen reports the total number between 20,000 and 30,000 diskettes. See Cohen, *A Short Course on Computer Viruses,* p. 50. Jan Hruska puts the number at 20,000. See Hruska, *Computer Viruses and Anti-Virus Warfare,* p. 20.

17. Philip Fites et al., *The Computer Virus Crisis,* p. 46.

rather ineffective form of extortion because users of the disk were required to mail payment of $189 (for a limited license) or $378 (for a lifetime license) to a post office box in Panama.

The virus author was eventually discovered to be an American named Joseph Popp who was extradited to Britain in February 1991 to face charges but was eventually dismissed as being psychiatrically unfit to stand trial.[18] He was later found guilty in absentia by an Italian court.

Other AIDS-related incidents include the early Apple II virus "Cyber-aids," the AIDS virus from 1989 that displays the message "Your computer now has AIDS" in large letters, followed a year later by the AIDS II virus that performs a similar infraction.

So here are two threat paradigms, terrorism and AIDS, which characterize the changing discursive position of computer viruses from the 1980s to 1990s. While the AIDS paradigm dominated in the late 1980s, by the late 1990s computer viruses would become weaponized and more closely resemble the terrorism paradigm.

The AIDS epidemic in the 1980s had a very specific discursive diagram. With AIDS, the victims became known, but the epidemic itself was unknown. There emerged a broad, immaterial social anxiety. The biological became dangerous and dirty. All sex acts became potentially deviant acts and therefore suspect.

But with terrorism there exists a different discursive diagram. With terror the victims are rarely known. Instead knowledge is focused on the threat itself—the strike happened here, at this time, with this weapon, by this group, and so on.

If AIDS is an invisible horror, then terror is an irrational horror. It confesses political demands one minute, then erases them the next (while the disease has *no* political demands). The state attacks terror with all available manpower, while it systematically ignores AIDS. Each shows a different exploitable flaw in protocological management and control.

While the shift in threat paradigms happened in the late 1980s for computer viruses, the transformation was long in coming. Consider the following three dates.

------

18. Hruska, *Computer Viruses and Anti-Virus Warfare,* p. 22.

In the 1960s in places like Bell Labs,[19] Xerox PARC and MIT scientists were known to play a game called Core War. In this game two self-replicating programs were released into a system. The programs battled over system resources and eventually one side came out on top. Whoever could write the best program would win.

These engineers were not virus writers, nor were they terrorists or criminals. Just the opposite, they prized creativity, technical innovation, and exploration. Core War was a fun way to generate such intellectual activity. The practice existed for several years unnoticed. "In college, before video games, we would amuse ourselves by posing programming exercises," said Ken Thompson, co-developer of the UNIX operating system, in 1983. "One of the favorites was to write the shortest self-reproducing program."[20] The engineer A. K. Dewdney recounts an early story at, I assume, Xerox PARC about a self-duplicating program called Creeper that infested the computer system and had to be brought under control by another program designed to neutralize it, Reaper.[21] Dewdney brought to life this battle scenario using his own gaming language called Redcode.

Jump ahead to 1988. At 5:01:59 p.m.[22] on November 2 Robert Morris, a 23-year-old graduate student at Cornell University and son of a prominent computer security engineer at the National Computer Security Center (a division of the NSA), released an email worm into the ARPAnet. This self-

---

19. A. K. Dewdney identifies a game called Darwin invented by M. Douglas McIlroy, head of the Computing Techniques Research Department at Bell Labs, and a program called Worm created by John Shoch (and Jon Hupp) of Xerox Palo Alto Research Center. See A. K. Dewdney, "Computer Recreations," *Scientific American,* March 1984, p. 22. For more on Shoch and Hupp, see "The Worm Programs," *Communications of the ACM,* March 1982. Many attribute the worm concept to the science fiction novel *Shockwave Rider* by John Brunner.
20. Ken Thompson, "Reflections on Trusting Trust," in *Computers Under Attack: Intruders, Worms, and Viruses,* ed. Peter Denning (New York: ACM, 1990), p. 98.
21. Dewdney, "Computer Recreations," p. 14.
22. Jon A. Rochlis and Mark W. Eichin, "With Microscope and Tweezers: The Worm from MIT's Perspective," in *Computers Under Attack: Intruders, Worms, and Viruses,* ed. Peter Denning (New York: ACM, 1990), p. 202. The precise time comes from analyzing the computer logs at Cornell University. Others suspect that the attack originated from a remote login at a MIT computer.

replicating program entered approximately 60,000[23] computers in the course of a few hours, infecting between 2,500 and 6,000 of them. While it is notoriously difficult to calculate such figures, some speculations put the damage caused by Morris's worm at over $10,000,000.

On July 26, 1989, he was indicted under the Computer Fraud and Abuse Act of 1986. After pleading innocent, in the spring of 1990 he was convicted and sentenced to three years' probation, fined $10,000, and told to perform four hundred hours of community service. Cornell expelled him, calling it "a juvenile act,"[24] while Morris's own dad labeled it simply "the work of a bored graduate student."[25]

While the media cited Morris's worm as "the largest assault ever on the nation's computers,"[26] the program was largely considered a sort of massive blunder, a chain reaction that spiraled out of control through negligence. As Bruce Sterling reports: "Morris said that his ingenious 'worm' program was meant to explore the Internet harmlessly, but due to bad programming, the worm replicated out of control."[27] This was a problem better solved by the geeks, not the FBI, thought many at the time. "I was scared," admitted Morris. "It seemed like the worm was going out of control."[28]

Morris's peers in the scientific community considered his prosecution unnecessary. As reported in *UNIX Today!*, only a quarter of those polled thought

23. Cohen, *A Short Course on Computer Viruses,* p. 49. The figure of 60,000 is also used by Spafford, who attributes it to the October 1988 IETF estimate for the total number of computers online at that time. See Eugene Spafford, "The Internet Worm Incident," in *Rogue Programs,* ed. Lance Hoffman (New York: Van Nostrand Reinhold, 1990), p. 203. Peter Denning's numbers are different. He writes that "[o]ver an eight-hour period it invaded between 2,500 and 3,000 VAX and Sun computers." See Denning, ed., *Computers Under Attack: Intruders, Worms, and Viruses* (New York: ACM, 1990), p. 191. This worm is generally called the RTM Worm after the initials of its author, or simply the Internet Worm.

24. From a Cornell University report cited in Ted Eisenberg et al., "The Cornell Commission: On Morris and the Worm," in *Computers Under Attack: Intruders, Worms, and Viruses* (New York: ACM, 1990), p. 253.

25. Cited in *The New York Times,* November 5, 1988, p. A1.

26. *The New York Times,* November 4, 1988, p. A1.

27. Bruce Sterling, *The Hacker Crackdown* (New York: Bantam, 1992), pp. 88–89.

28. Cited in *The New York Times,* January 19, 1990, p. A19.

Morris should go to prison, and, as the magazine testified, "most of those who said 'Yes' to the prison question added something like, 'only a minimum security prison—you know, like the Watergate people vacationed at.'"[29] Thus while not unnoticed, Morris's worm was characterized as a mistake, not an overt criminal act. Likewise his punishment was relatively lenient for someone convicted of such a massive infraction.

Ten years later, in 1999, after what was characterized as the largest Internet manhunt ever, a New Jersey resident named David Smith was prosecuted for creating Melissa, a macro virus that spreads using the Microsoft Outlook and Word programs. It reportedly infected over 100,000 computers worldwide and caused $80 million in damage (as assessed by the number of hours computer administrators took to clean up the virus). While Melissa was generally admitted to have been more of a nuisance than a real threat, Smith was treated as a hard criminal rather than a blundering geek. He pleaded guilty to ten years and a $150,000 fine.

With Smith, then, self-replicating programs flipped 180 degrees. The virus is now indicative of criminal wrongdoing. It has moved through its biological phase, characterized by the associations with AIDS, and effectively been weaponized. Moreover criminal blame is identified with the virus author himself who is thought of not simply as a criminal but as a cyberterrorist. A self-replicating program is no longer the hallmark of technical exploration, as it was in the early days, nor is it (nor was it ever) a canary in the coal mine warning of technical flaws in proprietary software, nor is it even *viral;* it is a weapon of mass destruction. From curious geek to cyberterrorist.

Cyberfeminism

Decades after programmers and pundits alike had safely agreed that computers were, at the end of the day, a decidedly *male* operation—for who else but the old boy's academo-military network had created the Internet, the

---

29. "Morris's Peers Return Verdicts: A Sampling of Opinion Concerning The Fate of the Internet Worm," in *Rogue Programs,* ed. Lance Hoffman (New York: Van Nostrand Reinhold, 1990), p. 104.

personal computer, cyberspace, viruses, video games, multimedia,[30] and so on—cultural critic Sadie Plant had this to say: "Hardware, software, wetware—before their beginnings and beyond their ends, women have been the simulators, assemblers, and programmers of the digital machines."[31] That the three occupations named here carry less clout than others one can imagine (Engineer, CEO, etc.) does not diminish the strength of Plant's argument: that computers are, and have always been, a technology of the female. Plant's coup is the unveiling of Ada Lovelace, a female protagonist drawn from computing prehistory. More on her later. Plant reaches beyond mythmaking—for what else can Lovelace be at this stage in the game—into a complex relationship between women and machines. This relationship, tied up in problematics surrounding identity, technology, and the body, is at the heart of the 1990s movement called cyberfeminism.

Cyberfeminism is a type of tactical media. It reflects on the totality of protocological command and control. Cyberfeminism adds a new dimension to the discussion begun in the previous sections on hackers and viruses, for this new strain deals with the negative space created within protocol through the injection of mutations, crashes, and viral code. With cyberfeminism, protocol becomes disturbed. Its course is altered and affected by the forces of randomness and corruption.

Indeed it is possible to think of cyberfeminism itself as a type of virus, a bug within the larger protocological network. Sadie Plant and others have identified Grace Hopper as the discoverer of the first computer bug. The bug was quite literally that, a moth caught in the innards of an early computing machine. The moth disrupted the normal functioning of the machine. Henceforth the term *bug* has been used to describe logical mistakes or glitches in computer code.

The computer bug, far from being an unwanted footnote in the history of computing, is in fact a space where some of the most interesting protocological

---

30. Packer and Jordan's 2001 anthology *Multimedia: From Wagner to Virtual Reality* is one of the more egregious examples. While their anthology is interesting, they essentially remove women from the history of multimedia, publishing in the first edition only one female author out of thirty-two texts, then adding a very short coda from Laurie Anderson in the "expanded" edition.
31. Sadie Plant, *Zeros and Ones* (New York: Doubleday, 1997), p. 37.

phenomena occur. Bugs, crashes, and viruses have always existed. (I argue in the last chapter that crashes actually define certain genres of contemporary Net art.) They are a sort of super-protocological mutation that can, at times, propel technology in interesting new ways.

"[O]ne of the guys [at ARPA] wrote a program called 'The Unknown Glitch,'" remembers computer pioneer Alan Kay, "which at random intervals would wake up, print out I AM THE UNKNOWN GLITCH. CATCH ME IF YOU CAN, and then it would relocate itself somewhere else in core memory, set a clock interrupt, and go back to sleep. There was no way to find it."[32] This Unknown Glitch was not anti-protocol by any means, for the very environment in which it thrived was the computer itself. Yet at the same time, the Glitch exists outside of the normal functionality of protocol. It is a liminal agent, at once inside protocol and outside its reach. This is the same status that cyberfeminism has now assumed.

The logical exploits described in chapter 5 also have immense implications in the realm of computer viruses. Computer viruses are, in essence, machines for the exploitation of logical flaws within a computer system. Viruses are not alive, at least not in any conventional sense of the word. But they are vital forms from the perspective of the "machinic phylum," that stratum of our material world populated by both physical and biological machines.

While they are often small, a virus's internal structure can be incredibly sophisticated. "What we have here is perhaps the most complex and refined malicious code in the history of virus writing," comments Eugene Kaspersky, Head of Company Anti-Virus Research Center, on the Hybris virus. "Firstly, it is defined by an extremely complex style of programming. Secondly, all the plugins are encrypted with very strong RSA 128-bit crypto-algorithm key. Thirdly, the components themselves give the virus writer the possibility to modify his creation 'in real time,' and in fact allow him to control infected computers worldwide."[33]

Viruses propagate themselves through weaknesses in the logical structure of computer code. Hackers often argue, in fact, that the logical weaknesses

---

32. Alan Kay, cited in Stewart Brand, "SPACEWAR: Fanatic Life and Symbolic Death Among the Computer Bums," *Rolling Stone,* December 7, 1972, p. 52.

33. Cited online at http://www.kaspersky.com.

themselves are the real problem, not the viruses that simply exploit the weakness. What is truly to blame, the water leaking from a bucket, or the hole in that bucket that allows the water to leak? Or, as the hacker magazine *2600* asked in response to the massive disturbance ($10 million of damage by professional estimates) caused by the "I Love You" virus: "How could it be possible to completely gloss over the fact that, once again, all of the problems were because of a gaping weakness in a program called Microsoft Outlook and that this is a lesson that should have been learned from the Melissa virus a year earlier?"[34] The affliction then becomes Microsoft Outlook—an anti-protocol application—not the "I Love You" virus.

(The addition of the virus deliberately complicates the issue, for if Microsoft Outlook were not monopolistic in the marketplace it would not as easily fall prey to infection. The greater saturation a particular application has, the higher likelihood that a virus will be able to spread. Either way, I draw a critical distinction in this book between proprietary software that happens to have a market monopoly and the universalism of a protocological technology.)

"The territory of cyberfeminism is large," write Faith Wilding and Critical Art Ensemble in their study of cyberfeminism. "It includes the objective arenas [of] cyberspace, institutions of industrial design, and institutions of education—that is, those arenas in which technological process is gendered in a manner that excludes women from access to the empowering points of techno-culture."[35]

History confirms this breadth. The first "Cyberfeminist Manifesto" appeared in the early nineties, written by a renegade group of Australian artists and activists calling themselves VNS Matrix. After this early rant, the cyberfeminist movement quickly grew on an international scale. On September 20, 1997, in Kassel, Germany, the First Cyberfeminist International met at Documenta X, an international exhibition of contemporary art.

Cyberfeminism in its very nature necessitates a participatory practice in which many lines of flight coexist. Yet several recurrent themes emerge,

---

34. "Madness," *2600* (Summer 2000), p. 5.

35. Faith Wilding and Critical Art Ensemble, "Notes on the Political Condition of Cyberfeminism," available online at http://www.obn.org/cfundef/condition.html.

among them the questions of *body* and *identity*. Like a computer virus, cyberfeminism exists to mutate and transform these questions, guiding them in new directions within the protocological sphere.

Sadie Plant and Allucquère Rosanne "Sandy" Stone are perhaps the two best entry points into contemporary cyberfeminist theory. It is Plant's view that technology is fundamentally female—not male as the legions of geeks, computer science teachers, and *Wired* magazine editors would have one believe. Stone, on the other hand, focuses on how virtual communities, far from being simple gathering places, actually *produce* things like bodies, identities, and spaces.

Like French feminist Luce Irigaray before her, Plant argues that patriarchal power structures, which have unequally favored men and male forms in society, should be made more equal through a process of revealing and valorizing overlooked female elements.

Her book *Zeros and Ones* turns on the story of Ada Lovelace, the world's first computer programmer. As assistant to Charles Babbage, Lovelace helped build early calculation machines that many consider critical to the prehistory of computer science. Championing Lovelace over Babbage, Plant's goal is to recuperate this lost female origin from within the history of technology.[36]

However, as her manifesto-like "Feminisations: Reflections on Women and Virtual Reality" shows, Plant wishes not to valorize some negative space created by patriarchy, but to unveil the always already feminine space of technology. This is ultimately a more powerful move, for instead of simply objecting to past inequalities, it reveals how many of those inequalities were unfounded. "Masculine identity has everything to lose from this new technics," prophesizes Plant. "The sperm count falls as the replicants stir and the meat learns how to learn for itself. Cybernetics is feminisation."[37]

The universality of protocol can give feminism something that it never had at its disposal, the obliteration of the masculine from beginning to end.

---

36. Ada Lovelace's influence has not been completely lost. Aside from her roles in various science fiction novels, there is the late, eponymous Web art site äda 'web (http://adaweb.walkerart.org) and Lynn Hershman Leeson's film *Conceiving Ada.*

37. Sadie Plant, "Feminisations: Reflections on Women and Virtual Reality," in *Clicking In,* ed. Lynn Hershman Leeson (Seattle: Bay Press, 1996), p. 37.

With inspiration from the VNS Matrix (self-styled "saboteurs of Big Daddy Mainframe"), Plant begins to define this pure feminine space and how it can inflect protocological space.

Zeros and Ones persuasively shows how women have always been inextricably involved with protocological technology. Using the telephone operator as an example, she argues that women have traditionally comprised the laboring core of networks of all kinds, particularly the telecommunications networks. From the power loom to typewriting, (even to the discovery of the computer bug), Plant categorizes technology as a fundamentally female object. Even the zero—the nothingness of binary code—has always been the 0-ther, the female.

On the writing of *Zeros and Ones,* Plant remembers: "When I started the book it was really to try and correct, what I thought was the great misconception at the moment about the relationship between women and computers in particular and technology in general. It seemed to me, that a lot of 'orthodox' feminist theory was still very technophobic."[38]

Technophobic she is not. Throughout Plant's book the intersection of woman and the protocological matrix is primary. This materializes itself historically in the matrix-based weaving processes of industrial power looms, in the predominantly female operators of phone networks, in the trope of the woman as computer programmer (Ada Lovelace, Grace Hopper) and in the weblike structure of cyberspace. Because of this history, Plant writes that technology threatens phallic control and is fundamentally a process of emasculation. "The matrix weaves itself in a future which has no place for historical man,"[39] says Plant. The digital provides a space of valences that exists outside of and potentially preempts patriarchal structures.

In other words, as protocol rises, patriarchy declines. As Plant describes it, "The introduction of binary code introduces a plane of equivalence which undermines the very foundations of a world in which male and female have played the roles of superstructure and material base."[40] In this model, binary

38. Available online at http://www.t0.or.at/sadie/intervw.htm.

39. Sadie Plant, "The Future Looms: Weaving Women and Cybernetics," in *Clicking In,* ed. Lynn Hershman Leeson (Seattle: Bay Press, 1996), p. 132.

40. Available online at http://www.t0.or.at/sadie/binary.htm.

code replaces what have traditionally been the producers of value, these being the phallus, the law, the father, and so on.

This process was described in chapter 1 as the movement from a structure based on hierarchy and centralized control to one based on horizontality and distributed control.

In Plant, technology is less a question of good or bad and more the possibility of an objective weakening of patriarchy (or its technological synonym, "propriety"). Cyberfeminism, for Plant, implies that an alliance "is being developed between women, machinery and the new technology that women are using."[41] And that new technology is, of course, protocol.

Held aloft, yet notably aloof from the cyberfeminist movement, is Sandy Stone, theorist of the history of cyberspace, desire, and the virtual body.[42] Stone's early essay "Will the Real Body Please Stand Up?"[43] helped set the stage for contemporary debates on the status of the body in virtual communities.

The place of the body is central to cyberfeminism. Yet in this analysis, bodies are not natural objects made of flesh and blood, but rather are complex intersections of materiality and meaning. Stone argues that binarisms such as nature/culture actually function logically as "a strategy for maintaining boundaries for political and economic ends, and thus a way of making meaning."[44] In this way, the insertion of the body into protocological space actually produces meaning through the articulation of differences between bodies and non-bodies, between spaces and non-spaces.

Like Foucault's rejection of the "repressive hypothesis" in Volume 1 of his influential *History of Sexuality,* Stone claims that new technologies are not transparent agents that remove issues of gender from view, but rather they proliferate the production and organization of gendered bodies in space. She

---

41. Available online at http://206.251.6.116/geekgirl/001stick/sadie/sadie.html.

42. A good place to start with Stone is her homestead at http://sandystone.com/. Although her published material is readily available, online users may access digitized versions of articles including "The Empire Strikes Back," "Violation & Virtuality," and "What Vampires Know" at http://eserver.org/gender/.

43. Allucquère Rosanne Stone, "Will the Real Body Please Stand Up?," in *Cyberspace: First Steps,* ed. Michael L. Benedikt (Cambridge: MIT Press, 1992).

44. Stone, "Will the Real Body Please Stand Up?," p. 102.

shows that the dominant spatial metaphor for interactions in virtual spaces is, simply enough, the metaphor of our daily physical, Cartesian space. And like our offline space, virtual spaces are inhabited by bodies with "complex erotic components."[45]

This working metaphor is of course totally arbitrary, as Stone points out, since there is nothing in the logic of digital networks that necessarily pre-structures itself as Cartesian, or body-based, or desiring. On the contrary, digital networks are non-Cartesian, are bodyless, and have little connection to the movements of human desire. Through the introduction of tactical protocols, which are always negotiated and agreed to in advance by all participants, digital networks *become* Cartesian, body-based, desiring, and so on. Cyberfeminism is the tactical process by which this reification will be refashioned.

Stone shows that communications technology is conventionally thought of as "1) an apparatus for the production of community . . . 2) an apparatus for the production of body . . . [and] 3) a mediating [agent] between bodies and selves . . . i.e., interfaces."[46] Protocological space is imagined as a prosthesis, as an enormous extension of one's physical body, and through this giant phantom limb (the Net) one interacts with other virtual bodies.

Participants in online communities like the object-oriented social spaces called MOOs "learn to delegate their agencies to body representatives [avatars] that exist in imaginal spaces contiguously with representatives of other individuals."[47] The creators of one of the most popular MOOs, LambdaMOO, describe this relationship of bodies in social terms: "LambdaMOO is a new kind of society, where thousands of people voluntarily come together from all over the world."[48] As Stone and others show, a participatory social practice (i.e., community) based on an imagined ether-scape of desiring and interacting bodies (i.e., protocol) is basic to how one conceptualizes digital spaces.

Cyberfeminist pioneers VNS Matrix provide the frontline guerrilla tactics for Stone and Plant's theoretical efforts. VNS Matrix emerged from

---

45. Stone, "Will the Real Body Please Stand Up?," p. 105.

46. Allucquère Rosanne Stone, *The War of Desire and Technology at the Close of the Machine Age* (Cambridge: MIT Press, 1995), p. 89.

47. Stone, *The War of Desire and Technology at the Close of the Machine Age,* p. 121.

48. LambdaMOO (telnet://lambda.moo.mud.org:8888).

Adelaide, Australia, in the summer of 1991. Francesca da Rimini (also known as Gashgirl and/or Doll Yoko) gives her story of how it all started:

Like all good coagulating stories it starts with slime, and maybe ends with blood. I live on the edge of the Australian desert in a small town of lies and whispers with a palpable palpitating underbelly . . . It was the summer of 91. Definitely not the summer of love. We were four girls. We were hot and bored and poor (for me not much has changed, except I am no longer bored). We decided to try and crack the porn cartel with some chick porn. We made some images on stolen computers, Beg, Bitch, Fallen, Snatch. We decided it was more fun playing with computers than end-lessly scanning our pussies and so Velvet Downunder morphed into VNS Matrix.[49]

VNS Matrix are Josephine Starrs, Julianne Pierce, Francesca da Rimini and Virginia Barratt,[50] who have perpetrated a series of cyberfeminist interven-tions including a "bad code" anti-video game targeted at girls (or at least not targeted at 14-year-old boys). Da Rimini (using the pseudonym Doll Yoko) writes, "cyberfeminism/s has become the field from which i work, from which multiple lines of flight errupt anarchically, generating dialogues, relations, conceptual and physical objects."[51]

The original VNS Matrix Cyberfeminist Manifesto effectively captures her sentiment:

we are the modern cunt

positive anti reason

unbounded unleashed unforgiving

we see art with our cunt we make art with our cunt

we believe in jouissance madness holiness and poetry

we are the virus of the new world disorder

rupturing the symbolic from within

saboteurs of big daddy mainframe

the clitoris is a direct line to the matrix

---

49. Available online at http://www.thing.net/~rdom/janrev97.01.html.

50. Available online at http://sysx.apana.org.au/artists/vns/.

51. Available online at http://sysx.apana.org.au/artists/vns/.

VNS MATRIX
terminators of the moral code
mercenaries of slime
go down on the altar of abjection
sucking the visceral temple we speak in tongues
infiltrating disrupting disseminating
corrupting the discourse
we are the future cunt[52]

Its slogan, "the clitoris is a direct line to the matrix," is meant to highlight a fundamental material coexistence between the machine and the female body.

Originally ignorant of the work of Sadie Plant, VNS Matrix built its own praxis centered on women and technology. Pierce notes, "at the same time as we started using the concept of cyberfeminism, it also began to appear in other parts of the world. It was like a spontaneous meme which emerged at around the same time, as a response to ideas like 'cyberpunk' which were popular at the time. Since then the meme has spread rapidly and is certainly an idea which has been embraced by many women who are engaged with techno theory and practice."[53]

Pierce notes that cyberfeminists have never been anti-protocol, but rather use protocological machines as an integral part of their political action, art, and writing. Da Rimini (writing as Doll Yoko) posted in June 1997, to the *Nettime* email list[54] that "as artists, [VNS Matrix] were serious bout usin strategies like irony 'n inversion of cultural stereotypes to raise some of the many issues around women and technology . . . access . . . education . . . jobs . . . portrayal of girls/chix/women in popular/games culture etc etc."[55] Da Rimini's writing style is typical of the VNS Matrix brand of cyberfeminism, a crude, confrontational liberationist politics for women in the digital matrix.

---

52. Available online at http://www.t0.or.at/dolores/manifesto/vnstoc.htm.

53. Available online at http://web.aec.at/www-ars/matrix.html.

54. *Nettime* is an email community devoted to "net criticism, collaborative text filtering and cultural politics of the nets." More information is available online at http://www.nettime.org.

55. Francesca da Rimini (as Doll Yoko), "bossy cunts online," *Nettime,* June 18, 1997.

Here are a few questions and answers I was able to pose to the VNS Matrix:

As part of VNS Matrix you helped coin the term "cyberfeminist." It seems that term had a rather short life—maybe 1991–1998? Do you consider this term dated or still relevant?
Josephine Starrs: I think cyberfeminism will go down in history with the other great avant-garde movements such as dadaism, surrealism and the situationists.
Francesca da Rimini: There are a number of chix in australia and the states and europe who are now identifying as "cyberfeminists" and exploring various philosophical, political and social implications of what it all might mean and do. I was deeply engaged with this debate in the early 90s, and as a member of VNS Matrix helped to make quite a prolific body of artwork and narrative texts which played with some of the issues surrounding the relations of gender and technology. but that was then, and now I've moved on to explore other fields of enquiry and media activism. But of course I still identify as a feminist, if not a cyberfeminist.

Throughout all of cyberfeminist theory the theme of bodies and identities dominates. As one essay notes, "Bodies generally are all the rage on the Net—whether they are obsolete, cyborg, techno, porno, erotic, morphed, recombined, phantom, or viral."[56] Indeed, much of the focus on bodies stems from the process of forgetting the body (or trying to forget about forgetting the body!).

As Stone and others have written, the advent of cyberspace is the story of bodies migrating and morphing into new contexts. In fact, Lynn Hershman Leeson goes so far as to claim that "new [Web] users are forming the largest immigration in history"[57]—a powerful idea to keep in mind, that computer use could possibly constitute a real *immigration* of bodies (from the offline to the online).

Cyberfeminism aims to exorcise the essentialized, uninterrogated female body (brought into existence as a by-product of the protocological revolution) through a complex process of revalorization and rebuilding.

---

56. Faith Wilding and Critical Art Ensemble, "Notes on the Political Condition of Cyberfeminism."
57. Lynn Hershman Leeson, "Romancing the Anti-Body: Lust and Longing in (Cyber)space," in *Clicking In,* ed. Lynn Hershman Leeson (Seattle: Bay Press, 1996), p. 328.

The Cartesian subject is no longer relevant here, as Plant explains:

Basically the two positions that are established at the minute are either that you talk about disembodiment or you talk about embodiment. Either you're out of the body in some stratospheric zone or you're in the organism. I think that neither of those are correct. When people talk about getting out of the body they are still assuming that there is some kind of great transcendent space like heaven for the soul, or something non-material at any rate, to occupy. And as far as I'm concerned that isn't there. The universe isn't like that, it's a material process not some sort of idealist construction. So you can't get out of matter, that's the crucial thing. But you can get out of the confining organization of matter which is shaped into things and of course, organisms. The organism is literally organized around its organs, the vocabulary says it all really.[58]

Contemporary cyberfeminist cultural production, including VNS Matrix's self-described "cunt art," follows Plant's guideline to the letter.

Like Fluxus artist Shigeko Kubota's 1965 performance "Vagina Painting" or Carolee Schneemann's "Interior Scroll" (1976), VNS Matrix focuses on a raw, fleshy, expressive use of the body.

Who are some interesting new media artists you've found that fit into the so-called "cyberfeminist" framework?
Josephine Starrs: I don't want to be exclusive . . . but my favourites have been Innen, from Hamburg, Bureau of Inverse Technology, from Australia, Mara Tralla from Estonia, Linda Dement and Zina Kaye from Australia, Rachel Baker from the UK, Rosie Cross for Geek Girl and of course there are some fabulous cyberfeminist theorists and activists.

Cyberfeminism is an attitude, not some lame revamp of seventies feminist consciousness-raising groups. I think cyberfeminists use the media and other institutions for their own subversive purposes. When VNS Matrix wrote the cyberfeminist manifesto for the 21st century and later the Bitch Mutant Manifesto, we were using language, performance, irony and humour to put flesh and filth into the machines and expose the gendered biases hardwired into computer culture.

---

58. Available online at http://www.altx.com/interviews/sadie.plant.html.

Plant, Stone, and the VNS Matrix are good allies for navigating the difficult questions that surround the tactical space of protocological networks. For, in essence, *they recognize that the "negotiatedness" of protocol, the fact that it is a universalism only achieved through prior negotiation and subsequent agreement, means that protocol can and will be different.*[59]

It matters little if gender disappears completely, or if it reemerges as a moniker of militancy. The political question is simply choosing how and when to inject change into protocol so that it aligns more closely with one's real desires about social life and how it ought better to be lived. This is the essence of tactical media.

Conflicting Diagrams

> Netwar is about the Zapatistas more than the Fidelistas, Hamas more than the Palestine Liberation Organization (PLO), the American Christian Patriot movement more than the Ku Klux Klan, and the Asian Triads more than the Costa Nostra.
>
> — JOHN ARQUILLA AND DAVID RONFELDT, *Networks and Netwars*

Arquilla and Ronfeldt coined the term *netwar,* which they define as "an emerging mode of conflict (and crime) at societal levels, short of traditional military warfare, in which the protagonists use network forms of organization and related doctrines, strategies, and technologies attuned to the information age."[60]

Throughout the years new diagrams (also called graphs or organizational designs) have appeared as solutions or threats to existing ones. Bureaucracy is a diagram. Hierarchy is one too, as is peer-to-peer. Designs come and go,

---

59. As Ben Braddock (Dustin Hoffman) says in the beginning of *The Graduate* about his future: "I want it to be . . . different."

60. Arquilla and Ronfeldt, *Networks and Netwars,* p. 6. A similar litany from 1996 reads: "netwar is about Hamas more than the PLO, Mexico's Zapatistas more than Cuba's Fidelistas, the Christian Identity Movement more than the Ku Klux Klan, the Asian Triads more than the Sicilian Mafia, and Chicago's Gangsta Disciples more than the Al Capone Gang." See John Arquilla and David Ronfeldt, *The Advent of Netwar* (Santa Monica: Rand, 1996), p. 5.

serving as useful asset managers at one historical moment, then disappearing, or perhaps fading only to reemerge later as useful again. The Cold War was synonymous with a specific military diagram—bilateral symmetry, mutual assured destruction (MAD), massiveness, might, containment, deterrence, negotiation; the war against drugs has a different diagram—multiplicity, specificity, law and criminality, personal fear, public awareness.

This book is largely about one specific diagram, or organizational design, called distribution, and its approximate relationship in a larger historical transformation involving digital computers and ultimately the control mechanism called protocol.[61]

In this diagramatic narrative it is possible to pick sides and describe one diagram as the protagonist and another as the antagonist. Thus the rhizome is thought to be the solution to the tree,[62] the wildcat strike the solution to the boss's control, Toyotism[63] the solution to institutional bureaucracy, and so on. Alternately, terrorism is thought to be the only real threat to state power, the homeless punk rocker a threat to sedentary domesticity, the guerrilla a threat to the war machine, the temporary autonomous zone a threat to hegemonic culture, and so on.

This type of conflict is in fact a conflict between different social structures, for the terrorist threatens not only through fear and violence, but specifically through the use of a cellular organizational structure, a distributed network of secretive combatants, rather than a centralized organizational structure employed by the police and other state institutions. Terrorism is a sign that we are in a transitional moment in history. (Could there ever be anything else?) It signals that historical actors are not in a relationship of equilibrium, but are instead grossly mismatched.

---

61. This is not a monolithic control mechanism, of course. "The Internet is a large machine," writes Andreas Broeckmann. "This machine has its own, heterogeneous topology, it is fractured and repetitive, incomplete, expanding and contracting" ("Networked Agencies," available online at http://www.v2.nl/~andreas/texts/1998/networkedagency-en.html).

62. This is Deleuze and Guattari's realization in *A Thousand Plateaus.*

63. For an interesting description of Toyotism, see Manuel Castells, *The Rise of the Network Society* (Oxford: Blackwell, 1996), pp. 157–160.

Armies facing off in *Spartacus* (1960)

Conflicting Diagrams

In recent decades the primary conflict between organizational diagrams has been
between hierarchies and networks: the asymmetrical conflicts of guerrilla warfare
terrorism, and so on. But what happens when the powers-that-be get smart and
actually evolve into networked power (something that has already taken place in
some instances)? In the future we are likely to witness this general shift, downwar
into a new bilateral organizational conflict of networks fighting networks.

Bugs swarming in *Starship Troopers* (1997)

It is often observed that, due largely to the original comments of net-working pioneer Paul Baran, the Internet was invented to avoid certain vulnerabilities of nuclear attack. In Baran's original vision, the organizational design of the Internet involved a high degree of redundancy, such that destruction of a part of the network would not threaten the viability of the network as a whole. After World War II, strategists called for moving industrial targets outside urban cores in a direct response to fears of nuclear attack. Peter Galison calls this dispersion the "constant vigilance against the re-creation of new centers."[64] These are the same centers that Baran derided as an "Achilles' heel"[65] and that he longed to purge from the telecommunications network.

"City by city, country by country, the bomb helped drive dispersion,"[66] Galison continues, highlighting the power of the A-bomb to drive the push toward distribution in urban planning. Whereas the destruction of a fleet of Abrams tanks would certainly impinge upon army battlefield maneuvers, the destruction of a rack of Cisco routers would do little to slow down broader network communications. Internet traffic would simply find a new route, thus circumventing the downed machines.[67]

---

64. Peter Galison, "War against the Center," *Grey Room* 4, Summer 2001, p. 20.

65. Baran writes: "The weakest spot in assuring a second strike capability was in the lack of reliable communications. At the time we didn't know how to build a communication system that could survive even collateral damage by enemy weapons. Rand determined through computer simulations that the AT&T Long Lines telephone system, that carried essentially all the Nation's military communications, would be cut apart by relatively minor physical damage. While essentially all of the links and the nodes of the telephone system would survive, a few critical points of this very highly centralized analog telephone system would be destroyed by collateral damage alone by missiles directed at air bases and collapse like a house of card." See Paul Baran, Electrical Engineer, an oral history conducted in 1999 by David Hochfelder, IEEE History Center, Rutgers University, New Brunswick, NJ, USA.

66. Galison, "War against the Center," p. 25.

67. *New Yorker* writer Peter Boyer reports that DARPA is in fact rethinking this opposition by designing a distributed tank, "a tank whose principal components, such as guns and sensors, are mounted on separate vehicles that would be controlled remotely by a soldier in yet another command vehicle." See "A Different War," *The New Yorker,* July 1, 2002, p. 61. This is what the military calls Future Combat Systems (FCS), an initiative developed by DARPA for the

(In this way, destruction must be performed absolutely, or not at all. "The only way to stop Gnutella," comments WiredPlanet CEO Thomas Hale on the popular file sharing protocol, "is to turn off the Internet."[68] And this is shown earlier in my examination of protocol's high penalties levied against deviation. One is completely compatible with a protocol, or not at all.)

Thus the Internet can survive attacks not because it is stronger than the opposition, but precisely because it is weaker. The Internet has a different diagram than a nuclear attack does; it is *in a different shape.* And that new shape happens to be immune to the older.

All the words used to describe the World Trade Center after the attacks of September 11, 2001, revealed its design vulnerabilities vis-à-vis terrorists: It was a tower, a center, an icon, a pillar, a hub. Conversely, terrorists are always described with a different vocabulary: They are cellular, networked, modular, and nimble. Groups like Al Qaeda specifically promote a modular, distributed structure based on small autonomous groups. They write that new recruits "should not know one another," and that training sessions should be limited to "7–10 individuals." They describe their security strategies as "creative" and "flexible."[69]

This is indicative of two conflicting diagrams. The first diagram is based on the strategic massing of power and control, while the second diagram is based on the distribution of power into small, autonomous enclaves. "The architecture of the World Trade Center owed more to the centralized layout of Versailles than the dispersed architecture of the Internet," wrote Jon Ippolito after the attacks. "New York's resilience derives from the interconnections it fosters among its vibrant and heterogeneous inhabitants. It is in decentralized structures that promote such communal networks, rather than in reinforced steel, that we will find the architecture of survival."[70] In the past the war against terrorism resembled the war in Vietnam, or the war

---

U.S. Army. It is described as "flexible" and "network-centric." I am grateful to Jason Spingarn-Koff for bringing FCS to my attention.

68. Cited in Gene Kan, "Gnutella," in *Peer-to-Peer: Harnessing the Power of Disruptive Technologies,* ed. Andy Oram (Sebastopol: O'Reilly, 2001), p. 99.

69. See *The al-Qaeda Documents: Vol. 1* (Alexandria, VA: Tempest, 2002), pp. 50, 62.

70. Jon Ippolito, "Don't Blame the Internet," *Washington Post,* September 29, 2001, p. A27.

Contagion

Early on, computer viruses were compared to biological contagion. In that early mode, the virus itself was the epidemic. Later the discourse on viruses turned toward weaponization and terrorism. Today the virus author is the epidemic, not the virus—the moral evaluation of the computer virus is eclipsed today by a search for its author, who is prosecuted as a terrorist.

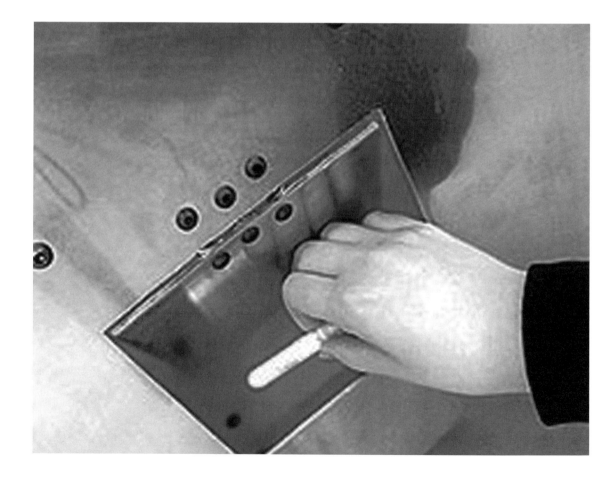

Experimental Interaction Unit (www.eiu.org), *Dispersion* (1999)

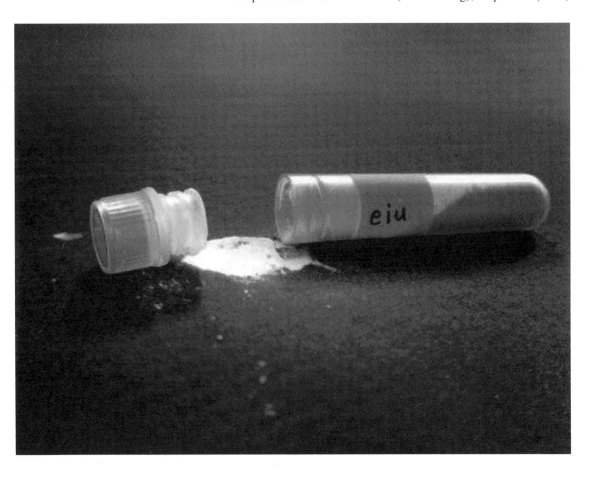

against drugs—conflicts between a central power and an elusive network. It did not resemble the Gulf War, or World War II, or other conflicts between states.

"As an environment for military conflict," *The New York Times* reported, "Afghanistan is virtually impervious[71] to American power." (In addition to the stymied U.S. attempt to rout Al Qaeda post-September 11, the failed Soviet occupation in the years following the 1978 coup is a perfect example of grossly mismatched organizational designs.) Being "impervious" to American power today is no small feat.

The category shift that defines the difference between state power and guerilla force shows that through a new diagram, guerillas, terrorists, and the like can gain a foothold against their opposition. But as Ippolito points out, this should be our category shift too, for anti-terror survival strategies will arise not from a renewed massing of power on the American side, but precisely from a distributed (or to use his less precise term, decentralized) diagram. Heterogeneity, distribution, and communalism are all features of this new diagrammatic solution.

In short, *the current global crisis is one between centralized, hierarchical powers and distributed, horizontal networks.* John Arquilla and David Ronfeldt, two researchers at the Rand Corporation who have written extensively on the hierarchy-network conflict, offer a few propositions for thinking about future policy:

---

71. Wanting instead American invulnerability to Soviet nuclear power, in 1964 Paul Baran writes that "we can still design systems in which system destruction requires the enemy to pay the price of destroying n of n [communication] stations. If n is made sufficiently large, it can be shown that highly survivable system structures can be built—even in the thermonuclear era." See Paul Baran, *On Distributed Communications: 1. Introduction to Distributed Communications Networks* (Santa Monica, CA: Rand, 1964), p. 16. Baran's point here is that destruction of a network is an all-or-nothing game. One must destroy all nodes, not simply take out a few key hubs. But the opposite is not true. A network needs only to destroy a single hub within a hierarchical power to score a dramatic triumph. Thus, Baran's advice to the American military was to become network-like. And once it did the nuclear threat was no longer a catastrophic threat to communications and mobility (but remains, of course, a catastrophic threat to human life, material resources, and so on).

- Hierarchies have a difficult time fighting networks. . . .
- It takes networks to fight networks. . . .
- Whoever masters the network form first and best will gain major advantages.[72]

These comments are incredibly helpful for thinking about tactical media and the role of today's political actor. It gives subcultures reason to rethink their strategies vis-à-vis the mainstream. It forces one to rethink the techniques of the terrorist. It also raises many questions, including what happens when "the powers that be" actually evolve into networked power (which is already the case in many sectors).

In recent decades the primary conflict between organizational designs has been between hierarchies and networks, an asymmetrical war. However, in the future the world is likely to experience a general shift downward into a new bilateral organizational conflict—networks fighting networks.

"Bureaucracy lies at the root of our military weakness," wrote advocates of military reform in the mid-eighties. "The bureaucratic model is inherently contradictory to the nature of war, and no military that is a bureaucracy can produce military excellence."[73]

While the change to a new unbureaucratic military is on the drawing board, the future network-centric military—an unsettling notion to say the least—is still a ways away. Nevertheless networks of control have invaded

---

72. Arquilla and Ronfeldt, *Networks and Netwars,* p. 15, emphasis removed from original. Contrast this line of thinking with that of Secretary of Defense Robert McNamara in the 1960s, whom Senator Gary Hart described as advocating "more *centralized* management in the Pentagon." See Gary Hart and William Lind, *America Can Win* (Bethesda, MD: Adler & Adler, 1986), p. 14. Or contrast it in the current milieu with the Powell Doctrine, named after four-star general and Secretary of State Colin Powell, which states that any American military action should have the following: clearly stated objectives, an exit strategy, the ability to use overwhelming force, and vital strategic interests at stake. This type of thinking is more in line with a modernist, Clausewitzian theory of military strategy: that force will be overcome by greater force, that conflict should be a goal-oriented act rather than one of continuance, that conflict is waged by state actors, and so on.

73. Hart and Lind, *America Can Win,* pp. 240, 249.

our life in other ways, in the form of the ubiquitous surveillance, biological informatization, and other techniques discussed in chapter 3.

The dilemma, then, is that while hierarchy and centralization are almost certainly politically tainted due to their historical association with fascism and other abuses, networks are both bad and good. Drug cartels, terror groups, black hat hacker crews, and other denizens of the underworld all take advantage of networked organizational designs because they offer effective mobility and disguise. But more and more one witnesses the advent of networked organizational design in corporate management techniques, manufacturing supply chains, advertisement campaigns, and other novelties of the ruling class, as well as all the familiar grassroots activist groups who have long used network structures to their advantage.

In a sense, networks have been vilified simply because the terrorists, pirates, and anarchists made them notorious, not because of any negative quality of the organizational diagram itself. In fact, positive libratory movements have been capitalizing on network design protocols for decades if not centuries. The section on the rhizome in *A Thousand Plateaus* is one of literature's most poignant adorations of the network diagram.

It has been the goal of this chapter to illuminate a few of these networked designs and how they manifest themselves as *tactical effects* within the media's various network-based struggles. As the section on viruses (or chapter 5 on hacking) showed, these struggles can be lost. Or as in the case of the end-to-end design strategy of the Internet's core protocols, or cyberfeminism, or the free software movement, they can be won (won in specific places at specific times).

These tactical effects are allegorical indices that point out the flaws in protocological and proprietary command and control. The goal is not to destroy technology in some neo-Luddite delusion, but to push it into a state of hypertrophy, further than it is meant to go. Then, in its injured, sore, and unguarded condition, technology may be sculpted anew into something better, something in closer agreement with the real wants and desires of its users. This is the goal of tactical media.

Internet Art

I feel it's time now to give a light on the origin of the term—
"net.art." Actually, it's a readymade. In December 1995 [Slovenian
artist] Vuk Cosic got a message, sent via anonymous mailer. Because of incompatibility of software, the opened text appeared to
be practically unreadable ascii abracadabra. The only fragment of it
that made any sense looked something like:

[. . .] J8~g#!\;Net. Art{-^s1 [. . .]

—ALEXEI SHULGIN, *Nettime*

In part III of this book, I have been examining the future of protocol. This future arrives through both successes and failures, and indeed failures that are also successes and successes that are also failures.[1] My previous discussion of hacking and tactical media shows that the advent and assumption of certain techno-subcultures both inaugurate the new protocological era and act to "crash" it by muddying its waters, jumping its fences, and generally monkey-wrenching with the new protocological systems of control.

Much of my analysis in preceding chapters focused on *form,* with the assumption that a revolutionary critique of the protocological media is simply a critique of their formal qualities: Determine a nonoppressive form and an emancipated media will follow. And indeed this is the main goal of media liberation theorists like Enzensberger.

The philosophy of formal critique later became a central problematic for many in the video movements of the 1960s and 1970s. What was at stake for video was the idea of specificity. For, the argument goes, if video is in fact a *specific* medium with its own practices and formal qualities, then it may distance itself from less desirable media such as television.

As museum director David Ross notes, "Video art has continually benefited from its inherently radical character. . . . [I]t has always been associated with the concepts of superindependent alternatives to the hegemony of commercial television."[2] Curator John Hanhardt agrees, writing that video was formed by "its opposition to the dominant institution of commercial television."[3]

Epigraphs: Alexei Shulgin, *Nettime,* March 18, 1997, cited in *Net_condition: Art and Global Media,* ed. Peter Weibel and Timothy Druckrey (Cambridge: MIT Press, 2001), p. 25. Vuk Ćosić, cited in Tilman Baumgärtel, "The Materiality of Test," Dec. 22, 1997, available online at http://www.rewired.com/97/1922.html.

1. William Morris's fantastic aphorism, which appears as the epigraph to Hardt and Negri's *Empire,* reads: "Men fight and lose the battle, and the thing that they fought for comes about in spite of their defeat, and then it turns out not to be what they meant, and other men have to fight for what they meant under another name." See Michael Hardt and Antonio Negri, *Empire* (Cambridge: Harvard University Press, 2000).

2. David Ross, "Forward," in *Illuminating Video,* ed. Doug Hall and Sally Jo Fifer (New York: Aperture, 1990), p. 10.

3. John Hanhardt, "Dé-collage/Collage: Notes Toward a Reexamination of the Origins of Video Art," in *Illuminating Video,* ed. Doug Hall and Sally Jo Fifer (New York: Aperture, 1990), p. 71.

Television was opposed for several reasons, including its centralized broadcast structure, its prohibitive costs, and its near total control by commercial interests. Thus, video is less a critical method than a critical *practice;* its mere existence is its attempt at critique.

It was more difficult for video artists to distance themselves from television than from film, for on the one hand the formal differences between film and video are manifest (magnetic tape versus celluloid, viewed on a monitor versus on a screen, low resolution versus high resolution, etc.), while on the other hand the differences between video and television are largely structural (individual versus commercial, local production/viewing versus large-scale production and broadcast, etc.).

Derrida offers an intriguing commentary on the question of video and its specificity as a medium. In doing so, he both empties it of its previous political content and injects it with a new utopian sensibility. After attacking video as having no essential unity or specificity, Derrida writes, in typically elliptical fashion, that "one never sees a new art, one thinks one sees it; but a 'new art,' as people say a little loosely, may be recognized by the fact that it is not recognized."[4] Thus, a truly subversive art form would, in fact, be invisible. The moment video is seen as art, it is divorced from its "newness."

Then, in a rare positive thrust, Derrida begins to map the terrain for a *radically* new type of video, what he describes as the *"possibility* that . . . is called video."[5] It is "vigilant" and "unpredictable" and it brings with it "other social spaces, other modes of production, of 'representation,' archiving, reproducibility . . . [and] the chance for *a new aura.*"[6]

Let me suggest that the "new art" that Derrida calls for is not in fact video, but the new media art that has appeared over the last few decades with the arrival of digital computers.[7] New media art—which I would define as

4. Jacques Derrida, "Videor," in *Resolutions,* ed. Michael Renov and Erika Suderburg (Minneapolis: Minnesota University Press, 1996), p. 75.

5. Derrida, "Videor," p. 73, emphasis mine.

6. Derrida, "Videor," p. 77, emphasis mine.

7. At the end of the 1960s computers were only beginning to enter the art world, despite the fact that they had been in public use for over two decades. The organization Experiments in Art and Technology, founded in 1966, was a pioneer in this area, producing a series of perfor-

any contemporary art that uses new media technology—covers the fields of Internet art, CD-ROM, certain kinds of installation art, digital video, electronic games, Net radio, and so on. Internet art, more specifically, refers to any type of artistic practice within the global Internet, be it the World Wide Web, email, telnet, or any other such protocological technology. Further, as I argue in this chapter, a subgenre of Internet art has emerged since 1995 called "net.art." This subgenre refers to the low-tech aesthetic popularized by the *7-11* email list and artists like Jodi.[8]

Media critic Timothy Druckrey writes that the first recorded usage of this term was on the *Nettime* email list. In a message written by Russian artist Alexei Shulgin, a citation from Slovenian artist Vuk Ćosić mentions that the expression "net.art" was created by accidentally picking out two connected words in a corrupted, unreadable email message.[9] The first critical discussion

---

mances that year called "Nine Evenings." Gene Youngblood's *Expanded Cinema* from 1970 is often cited as the first book to address art and new technologies at any length. Other texts from that period address the question of machinic technology and begin to touch on the question of electronics or computers. See Jack Burnham's "Systems Esthetics," *Artforum,* September 1968; Grace Marmor Spruch's interesting "Report on a Symposium on Art and Science Held at the Massachusetts Institute of Technology, March 20–22, 1968," *Artforum,* January 1969; Robert Mallary's "Computer Sculpture: Six Levels of Cybernetics," *Artforum,* May 1969; Thelma R. Newman's interview with Les Levine, "The Artist Speaks: Les Levine," *Art in America,* November 1969; J. W. Burnham's "The Aesthetics of Intelligent Systems," in the Guggenheim Museum's *On the Future of Art* (New York: Viking, 1970). A 1969 exhibition at the London ICA called "Cybernetic Serendipity," plus two shows in 1970, "Information" at the New York Museum of Modern Art and "Software" at the Jewish Museum, are considered to be the first museum shows to deal directly with computer art. "Software" was reviewed widely in such publications as *Art International, Art News,* and *Artforum.* The journal *October,* often a barometer for artistic and intellectual trends, does not cover the media arts until 1985 with Raymond Bellour's "An Interview With Bill Viola," *October,* Fall 1985.

8. See http://www.7-11.org and http://www.jodi.org. Jodi is a collaboration between Dirk Paesmans and Joan Heemskerk.

9. See Timothy Druckrey's essay entitled "[. . .] J8~.g#!\;NET.ART{-s1[. . .," in *Net_condition: Art and Global Media,* ed. Peter Weibel and Timothy Druckrey (Cambridge: MIT Press, 2001), p. 25. Some consider Vuk Ćosić's description of the origin of the word "net.art" to be embellished or possibly apocryphal. It's important to point out also that the term was not actually coined in Alexei Shulgin's March 18, 1997, email to the *Nettime* list. The term was in

of net.art appeared around 1997, as Druckrey notes: "The first extended discussion of net.art appeared in *ZKP 4*,"[10] a broadside published in 1997 in Ljubljana by the *Nettime* email community. The *ZKP4* (the fourth in a series of "ZKP" publications from *Nettime*'s so-called Central Committee) had a print run of 10,000 copies and is also available online.[11] The term "net.art" was in common use by the winter of 1996–1997.

I argue in this chapter that the definition of Internet art has always been a tactical one, that Internet art doesn't simply mean using browsers and HTML, but instead is an aesthetic defined by its oppositional position vis-à-vis previous, often inadequate, forms of cultural production. While the history of film and video practice is important, the greatest struggle of Internet art has been to prove its autonomy as an artistic practice—in the same way that video longed to be different from television. Marshall McLuhan offered a useful insight in this context. He wrote that the content of every new medium is generally the previous medium. That is to say, as new media formats appear historically, they often appear as mere wrappers for older formats—a perfect example of the logic of protocol.

Only through distinct breaks with the past will a medium gain its own specificity. For instance, cinematic techniques during the primitive phase of filmmaking at the turn of the century were primarily holdovers from previous entertainment formats such as vaudeville. Many shots were staged in the manner of a theatrical performance, with the camera held in a stationary position (mimicking the imaginary theatergoer's point of view) opposite a two-dimensional tableau formation of actors. Only later did filmmakers begin to move the camera, and thus begin to experiment with a specifically cinematic method of representation.

In the case of the Internet, many have tried to make painting or video or even hypertext the content of Internet art, yet they are thwarted by several factors that are unique to the medium, what might be called the medium's

---

circulation on several email lists, including *Nettime* and *Rhizome,* prior to March 1997. For example, the first recorded use of the expression "net.art" on the *Rhizome* email list was by Vuk Ćosić in May, 1996 in his announcement for the gathering entitled "Net.art per se."

10. Druckrey, "[. . .] J8~.g#!\;NET.ART{-s1 [. . .," p. 25.

11. See http://www.nettime.org.

Web site specificity. Marina Gržinić has commented interestingly on this fact in her essay "Exposure Time, the Aura, and Telerobotics" where she argues that the very limitations of new media technology, what she describes as the "delays in transmission-time, busy signals from service providers, crashing web browsers,"[12] are what bring about its specificity as an artistic medium. Always at the margins of the art world, Internet art has massively disengaged itself from mainstream practices in order to find its own space. Following Gržinić, I suggest here that computer crashes, technical glitches, corrupted code, and otherwise degraded aesthetics are the key to this disengagement. They are the "tactical" qualities of Internet art's deep-seated desire to become specific to its own medium, for they are the moments when the medium itself shines through and becomes important.

Internet art emerged in a specific political context. The two dominant forces vis-à-vis modern computing were hacking, which predates Internet art by many years, and the rather recent invention (at least in its present incarnation) of tactical media, both of which I have discussed in previous chapters. As stated earlier, computer hacking was the first cultural practice to employ affected interaction with computers. Its superficial virtues are unsanctioned exploration and pure freedom of information. In its depoliticized form, hacking is simply curious exploration. However, in its politicized form hacking generally follows a libertarian philosophy: Freedom for all information, down with bureaucratic control, and get the cops/teachers/parents off our backs.

Here is The Mentor's "Hacker Manifesto" again:

We explore . . . and you call us criminals. We seek after knowledge . . . and you call us criminals. We exist without skin color, without nationality, without religious bias . . . and you call us criminals. . . . Yes, I am a criminal. My crime is that of curiosity. My crime is that of judging people by what they say and think, not what they look like. My crime is that of outsmarting you, something that you will never forgive me for. I am a hacker, and this is my manifesto.[13]

---

12. Marina Gržinić, "Exposure Time, the Aura, and Telerobotics," in *The Robot in the Garden,* ed. Ken Goldberg (Cambridge: MIT Press, 2000).

13. See http://www.iit.edu/~beberg/manifesto.html.

This type of rhetoric—"we explore . . . and you call us criminals, we seek after knowledge . . . and you call us criminals"—is common in hacker manifestos. Many on the Left have been disappointed with the political potential of hacking because of this libertarian, gee-whiz desire for freedom of information. Tactical media, on the other hand, is almost synonymous with the Left because it is driven almost exclusively by progressive politics.

These two worlds collided in September 1998 at the Ars Electronica Festival in Linz, Austria, when the Electronic Disturbance Theater (EDT) was criticized by the HEART group (Hackers for Electronic Arts). The dispute was sparked by a piece of software used by the EDT. The software, called *Floodnet*, uses a technique called a Distributed Denial of Service (DDoS) attack to stage political protests on the Internet. (The EDT has used *Floodnet* to stage dozens of these so-called virtual sit-ins in solidarity with the Mexican Zapatista movement.) In the same way that a real-world protest helps bring a certain political cause into the public eye, *Floodnet* is primarily a *visualization* tool, but for abstract networks rather than real world situations. It makes the Internet and the people in it more visible—and their political cause with them—by creating what EDT founder Ricardo Dominguez calls "disturbances" within protocol. Like many other examples of tactical media, *Floodnet* cannot be categorized as merely an art project or merely a political tool, but must be both at the same time. Its ability to aesthetically render the abstract space of protocological networks into a visible "disturbance" is precisely its value as both a political tool and a work of art.

The HEART hackers argued however that *Floodnet* should not be deployed because, by using the DDoS attacks to create disturbances on the Internet, it in essence limits access to information. Undesirable information, perhaps, but information nonetheless. Freedom of all information was more important to the HEART hackers than the political disturbances. Furthermore, the hackers suggested that *Floodnet* was technically flawed because it was relatively easy to defeat.

As I suggest in part I, the protocols that underlie the Internet are not politically neutral. They regulate physical media, sculpt cultural formations, and exercise political control. This fact helps one understand the difference of opinion between the hackers and the artists/activists. If the network itself is political from the start, then any artistic practice within that network must engage politics or feign ignorance.

Simple Net Art Diagram

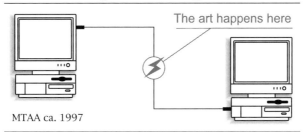

The art happens here

MTAA ca. 1997

Figure 7.1
"Simple Net Art Diagram" (Courtesy of M.River and T.Whid Art Associates; 1997.)

This argument is very similar to Blake Stimson's account of the origins of conceptual art when he argues that the elevated political climate of the 1960s was largely responsible for creating conceptualism as an artistic movement: "Conceptualism challenged the authority of the institutional apparatus framing its place in society and sought out other means for art to function in the world."[14] One must consider the network itself to be the "institutional apparatus" responsible for the birth of today's Internet artists (see figure 7.1).

Let me now take a closer look at Internet art by examining some of its specific aesthetic qualities. The Internet's early autonomous communities were the first space where pure network aesthetics (Web site specificity) emerged—email lists like *7-11, Nettime, recode, Rhizome,* and *Syndicate.*

Primitive signs were seen in early net.art projects, such as Alexei Shulgin's *Refresh,* an art project consisting of nothing but links between Web pages.[15] *Refresh* involves many different organizations working together, using many different computers all around the world. In *Refresh* a chain of Web pages is created. Each page is programmed to link automatically (on a 10-second delay) to the next Web page in the chain. Shulgin describes the project as "A Multi-Nodal Web-Surf-Create-Session for an Unspecified Number of Players."[16] Anyone can collaborate in the project by slipping his or her own page

---

14. Blake Stimson, "The Promise of Conceptual Art," in *Conceptual Art: A Critical Anthology,* ed. Alexander Alberro and Blake Stimson (Cambridge: MIT Press, 1999), p. xxxix.

15. See http://sunsite.cs.msu.su/wwwart/refresh.htm.

16. See http://sunsite.cs.msu.su/wwwart/refresh.htm.

into the link of refreshes. The user may load any Web page in the chain, and then watch as a new Web site appears every several seconds like a slide show.

In this way, *Refresh* was one of the first works to render the network in an artistic way—as a painter renders a landscape or a sculptor renders a physical form. The art exists "out there" in the network, not on any individual Web page in the chain. *Refresh* made visible a virtual network of collaboration that was not based on individual content. Shulgin's work spatializes the Web. It turns the Internet, and protocol with it, into a sculpture.

Journalist and cultural critic Tilman Baumgärtel articulates this self-referential quality of Internet art very clearly when he writes:

It has always been emphasized that the first and most important theme of Net art is the Internet itself. Net art addresses its own medium; it deals with the specific conditions the Internet offers. It explores the possibilities that arise from its taking place within this electronic network and is therefore "Net specific." *Net art plays with the protocols of the Internet, with its technical peculiarities.* It puts known or as yet undiscovered errors within the system to its own use. It deals creatively with software and with the rules software follows in order to work. It only has any meaning at all within its medium, the Internet.[17]

While Shulgin's work is highly conceptual, more formal work was also produced in this period. Perhaps the best example of formal work is from the European duo Jodi.[18] For several years Jodi has refined a formal style by making computers both the subject and content of their art making. Focusing specifically on those places where computers break down, Jodi derives a positive computer aesthetic by examining its negative, its point of collapse.

For example, in Jodi's work *404,*[19] which alludes to the Web's ubiquitous "file not found" 404 error code (which is built into Berners-Lee's HTTP protocol), the artists use the default fonts and simple colors available to primitive Web browsers. *404* is a collection of pages where users can post text

---

17. Tilman Baumgärtel, *net.art 2.0: New Materials towards Net art* (Nürnberg: Verlag für moderne Kunst Nürnberg, 2001), p. 24, emphasis mine.

18. See http://www.jodi.org.

19. See http://404.jodi.org.

messages and see what other users have written. But this simple bulletin board system becomes confused as the input text is pushed through various distorting filters before being added to the Web page for general viewing. The result is a rather curious collection of bathroom-wall scrawl that foregrounds the protocols of the Web page itself, rather than trying to cover over the technology with pleasing graphics or a deliberate design.

The 404 error code has also been used by other artists. Lisa Jevbratt's "Non-Site Gallery" opens up the dead end of the 404 error page. She transforms the 404 message into a generative doorway, where the requested page is generated on the fly, as if it had always existed for the user and was not the result of a mistake.

The 404 error code was also used in a more conceptual sense by the EDT. As part of its virtual sit-ins the EDT have created software that sends out Web requests for nonexistent Web pages on remote servers embedded with special messages—addresses in the form of www.server.com/__special_ message__. Since the Web pages do not exist on the remote server (and were never intended to exist), an error message is immediately generated by the server and returned to the EDT software.

However—and this is the trick—since Web servers record *all* traffic to their Web site including errors, the error acts like a Trojan horse and the "special message" is recorded in the remote server's log book along with the rest of its Web traffic. This accomplishes the difficult task of actually uploading a certain specified piece of information to the server of one's choice (albeit in a rather obscure, unthreatening location). As the messages pass from the protester to the protested site, a relationship is created between the local user and the remote server, like a type of virtual sculpture.

While the artwork may offer little aesthetic gratification, it has importance as a conceptual artwork. It moves the moment of art making outside the aesthetic realm and into the invisible space of protocols: Web addresses and server error messages.

As work from the EDT suggests, Internet conceptualism is often achieved through a spatialization of the Web. It turns protocol into a sculpture. As the Internet changes, expanding its complex digital mass, one sees that the Web itself is a type of art object—a basis for myriad artistic projects. It is a space in which the distinction between art and not art becomes harder and harder to see. It is a space that offers itself up *as art.*

The cluster of servers that make up the *Name.Space* alternative network—a web within the Web that uses a different, more flexible (not to mention cheaper and nonmonopolistic) addressing scheme—are a perfect example of this type of Internet conceptualism.[20] Control over Internet naming (DNS) is crucial for *Name.Space* founder Paul Garrin who considers it a type of poetic subversion to break out of the limitations of the com/edu/net addressing scheme for top-level domain names. *Name.Space* is a strategic intervention within the structure of the Web—art and politics are inseparable in this instance. Garrin calls the art project an "independent tactical network," with the goal of insuring that there will always be "a home for free media and alternative voices and visions on the ever changing internet."

The *Web Stalker*[21] is also a good example of the conceptual nature of Internet art. It is an alternate browser that offers a completely different interface for moving through pages on the Web. The *Web Stalker* takes the idea of the visual browser (e.g., Netscape Navigator or Internet Explorer) and turns it on its head. Instead of showing the art on the Web through interpreting HTML and displaying in-line images, it exhibits the Web itself *as art* through a making-visible of its latent structure. The user opens a Web address, then watches as the *Stalker* spits back the HTML source for that address. In a parallel window the *Web Stalker* exhaustively maps each page linked from that URL, exponentially enlarging the group of scanned pages and finally pushing an entire set of interlinked pages to the user. The pages are mapped in a deep, complex hypertextual relation.

The *Web Stalker* doesn't produce art but, in Matthew Fuller's words, "produces a *relationship* to art."[22] The *Stalker* slips into a new category, the "not-just-art" that exists when revolutionary thinking is supplemented by aesthetic production.

Let me now propose a simple periodization that will help readers understand Internet art practice from 1995 to the present. *Early Internet art—the highly conceptual phase known as "net.art"—is concerned primarily with the net-*

---

20. See http://name.space.xs2.net.

21. See http://www.backspace.org/iod.

22. Matthew Fuller, "A Means of Mutation," available online at http://bak.spc.org/iod/mutation.html.

work, while later Internet art—what can be called the corporate or commercial phase—has been concerned primarily with software. This is the consequence of a rather dramatic change in the nature of art making concurrent with the control societies and protocological media discussed throughout this book.

The first phase, net.art, is a dirty aesthetic deeply limited, but also facilitated, by the network. The network's primary limitation is the limitation on bandwidth (the speed at which data can travel), but other limitations also exist such as the primitive nature of simple network protocols like HTML. Because of this, one sees a type of art making that is a mapping of the network's technological limitations and failures—as the wasp is a map of the orchid on which it alights, to use Deleuze and Guattari's expression. Examples include Jodi, Olia Lialina, Heath Bunting, Alexei Shulgin, Vuk Ćosić, and many others. Net.art is a very exciting aesthetic, full of creativity and interesting conceptual moves.

Yet this first phase may already be coming to an end. Baumgärtel recently observed that it is "the end of an era. The first formative period of net culture seems to be over."[23] He is referring to a series of years from 1995 to 1999 when the genre of net.art was first developed. In this period, due to prominent technical constraints such as bandwidth and computer speed, many artists were forced to turn toward conceptual uses of the Internet that were not hindered by these technical constraints, or, in fact, *made these constrains the subject of the work.* All art media involve constraints, and through these constraints creativity is born. Net.art is low bandwidth through and through. This is visible in ASCII art, form art, HTML conceptualism—anything that can fit quickly and easily through a modem.

But this primary limitation has now begun to disappear. Today Internet art is much more influenced by the limitations of certain commercial contexts. These contexts can take many different forms, from commercial animation suites such as Flash, to the genre of video gaming (a fundamentally commercial genre), to the corporate aesthetic seen in the work of RTMark, Etoy, and others. My argument is aesthetic, not economic. Thus, it is not a

---

23. Tilman Baumgärtel, "Art on the Internet—The Rough Remix," in *README!* ed. Josephine Bosma et al. (New York: Autonomedia, 1999), p. 229.

question of "selling out" but rather of moving to a new artistic playing field. As computers and network bandwidth improved during the late 1990s, the primary physical reality that governed the aesthetic space of net.art began to fall away. Taking its place is the more commercial context of software, what may be seen as a new phase in Internet art. Let me consider these two phases in turn.

Internet Art as Art of the Network

All art up to now has been merely a substitute for the Internet.
—VUK ĆOSIĆ

Amid the proliferation of hi-tech graphic design, browser plug-ins, and special media applications that appeared in the years 1995–1999, many art Web sites ignored such technological improvements and instead concentrated on making a new kind of Web-specific art that focused on the Internet itself as art object and receptacle. Instead of scanning offline art and porting it over to the Internet or digitizing film and uploading it to a Web server (an unfortunate practice known as shovelware), artists like Jodi made art specifically for, and of, the Web.

Jodi love the look of raw computer code and use it often in their work; the duo love snapshots of computer desktops; they love the aesthetic of the computer crash. With a rather degraded and simplified aesthetic, Jodi's project entitled *day66*[24] typifies net.art. With illegible images stacking up in the background and prominent use of the Javascript "scroll" feature, the piece skids into view. Just as the page loads, it begins frantically to move, scrolling diagonally across the screen as if the user's operating system had been replaced by a massive conveyor belt.

While it may be easy for some to write off Jodi as so much hostile nonsense, a certain type of technological aesthetic present in their work is worth a second look. Past the full-screen blink tags, and past the wild animated GIF images, there is a keen interest in computer protocols themselves as the focal point and structuring framework for artistic production. No other style of net.art reflects so directly on the nature of the Web as a medium.

---

24. See http://www.jodi.org/day66/.

With *OSS*,[25] Jodi continued to explore the margins of computer programming. Yet here the duo began to stretch outward, past their normal creative purview. Issued on both CD-ROM and as a standalone application, *OSS* has the ability to mimic a computer operating system. (Strictly speaking, *OSS* is a hybrid piece that uses both Internet and non-Internet technologies; however, since it incorporates the network in interesting ways and is so intimately connected to the artists' overall aesthetic strategy, it is worth examining here in detail.) Once launched, *OSS* hijacks the computer and forbids it from functioning normally. It's "the digital equivalent of an aneurysm,"[26] writes Steven Johnson about the piece. Like an operating system, *OSS* controls the appearance and functionality of the entire visual environment including desktop and pull-down menus. Within this counterintuitive (if not frightening) interface, *OSS* presents the user with a series of abstract, computer-based aesthetic experiences, many of which continue the chaotic, "computer virus" style seen at the Jodi Web site. Using CD-ROM, however, Jodi is able to obtain a much more immersive effect. Images and shapes take over the entire screen, not simply within a single Internet browser window.

The *OSS* CD-ROM has four basic areas, each with cryptic names like "#Reset;" or "%20." These four basic areas plunge the user into different visual environments. A fifth area, the folder named "****," contains 255 small (6k) SimpleText pictures and folders. Each of these is represented by an icon. Dozens more icons spread over the desktop. As icons, they provide the visual raw materials for *OSS*'s original four areas.

One piece, "%20," takes the desktop at face value, then electrocutes it. The desktop begins to shake uncontrollably, then loses all vertical hold and slides ungracefully off the screen. The colors begin to modulate, and the screen flickers. Degradation of the desktop can be arrested somewhat by moving the mouse, or via keyboard commands.

---

25. See http://oss.jodi.org. The name "OSS" is a pun with a cluster of connotations, including the German "SS," the distress call "S.O.S.," as well as "operating system." In personal correspondence the artists claim, however, that it is merely a coincidence that OSS is also a reference to the American organization called the Office of Strategic Services, a predecessor to the Central Intelligence Agency.

26. Steven Johnson, *Emergence* (New York: Scribner, 2001), p. 175.

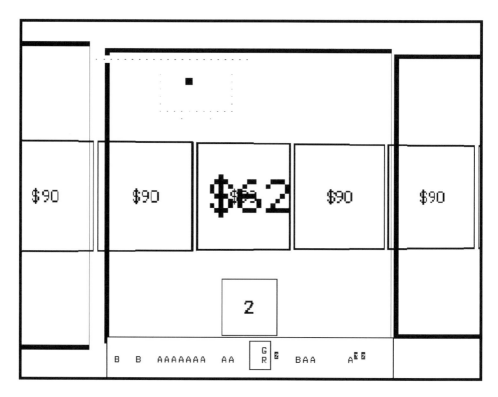

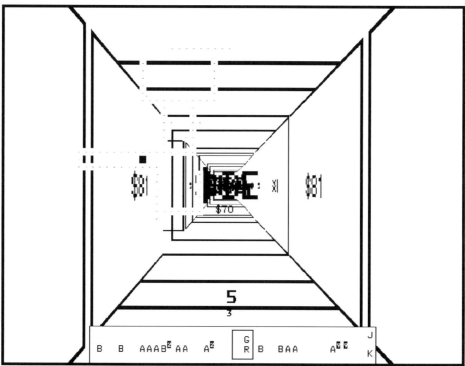

Jodi, *sod-b.gif* (2003)

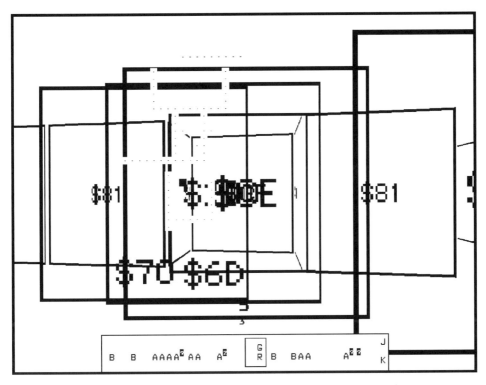

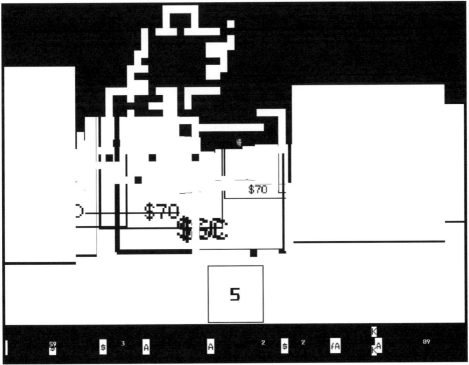

Jodi, *sod-b2.gif* (2003)

Another section of the work, called "#Reset;," resembles op art. Images scroll up and down the screen, moving so rapidly that new shapes begin to appear out of the interference patterns between shapes—like the spokes on a quickly moving wheel appearing to rotate backward through optical illusion.

The area called "**** ***" emulates the computer's desktop environment but reproduces it in horrible disarray: windows are spawned endlessly; the mouse draws a line as it moves, rather than performing its normal function as faithful point-and-click tool; the pull-down menu options are transformed into cryptic, useless ornaments. There seems to be no way out. Small hyphens in the pull-down menus allow the user to change the desktop background and mouse drawing color.

The "O00,0" environment is the least interesting. Playing what amounts to a frustrating game of pin the tail on the donkey, the user must click on a target "+" without being able to see the mouse pointer. Being blindfolded with a mouse is difficult indeed. The user may type "Control-Q" for quick escape. Several URLs appear at the top of the screen. Each of the 738 pin-the-tail targets is dutifully organized as a separate Web page in a folder at http://www.jodi.org/usemap/coords/. Exactly why is unclear.

OSS is abstract art for computers. In it, content itself has been completely subordinated to the sometimes jarring and pixelized topography of the computer operating system. Focusing specifically on those moments where computers break down (the crash, the bug, the glitch), Jodi discovers a new, autonomous aesthetic. That *OSS* is not strictly Internet-based does not preclude it from being included in the net.art genre, for the defining characteristic of net.art is a tactical relationship to protocols, not simple engagement with this or that technology.

Russian artist Olia Lialina has also worked in the net.art genre. In *Will-n-Testament,*[27] a Web project containing Lialina's will, the artist lists various pieces of digital property—her Internet art projects, images, pieces of writing—and assigns each to a friend or relative. Each benefactor is named. Each object is linked for public display. Lialina's corrections to the will are visible in blue ink. It is a very personal drama, the drama of her own death.

---

27. See http://will.teleportacia.org.

However, the interesting element vis-à-vis net.art lies outside this narrative structure. Each letter in the text of the will is in fact an image file. Instead of the letter "a," an image of an "a" is replaced, and so on for each letter of the alphabet. Since images load more slowly over the Internet than does text, the entire will takes several seconds to load fully on the computer screen. The art project is deliberately retarded, disabled by its own bloated size and limited bandwidth. Each letter loads at a different speed, causing the will to appear slowly in front of the user in a random sequence of letters. By making the download time part of the viewing experience, Lialina brings protocol itself directly into the art object.

Net.art's unique protocological characteristics are also seen in Shulgin's "Form Art" competition and exhibition.[28] Form Art refers to any Internet art piece that uses only the radio buttons, pull-down menus, and textboxes found in HTML forms. Self-consciously simplistic and technically restrained, form art uses HTML to explore and exploit new aesthetic possibilities. Shulgin's aesthetic is spur-of-the-moment, ephemeral, and totally contingent on a specific protocol (HTML). There is no depth to this work, rather there is an aesthetic of relationality, of machines talking to machines.

Heath Bunting, in projects such as _readme, has focused on a total dissolution of the art object into the network. _readme is similar to Lialina's *Will-n-Testament,* but transforms digitized text in a slightly different manner. After copying a randomly selected magazine article onto his Web page, Bunting modified the article so that each word of written text becomes a hyperlink to itself. For example, the word "is" links to www.is.com, "on" links to www.on.com, "together" links to www.together.com, and so on. The selection of links is not meaningful—some words have been bought as Internet addresses while other words remain inaccessible. As a Web page _readme is nothing but links to other places; it is an aestheticization of protocol as such.

In November 1998 at the Kunstlerhaus Bethanien in Berlin, Bunting created a very unique work of art for the "Net—Art—World: Reception Strategies and Problems" conference on net.art. Bunting had already gained a reputation in net.art circles as being somewhat aloof, a digital nomad who

---

28. See http://www.c3.hu/hyper3/form.

reputedly owned no possessions except for a single set of clothes and a CD-ROM that hung on a chain around his neck. Rumors had also circulated that Bunting, dissatisfied with harassment from the Euro-American art clique, had turned to making works of cyberterrorism, funded exclusively by rogue nations.

During his presentation in Berlin, Bunting stepped onstage and attempted to display a Web art project hosted on the Cuban domain www.castro.cu. While the audience waited for the art project to download and appear on the overhead screen, Bunting continued to talk about his other work. After a minute or two, the Web site request timed out and returned an error message. Embarrassed, Bunting quickly typed in another address under the www.castro.cu domain, hoping to save face in front of the waiting audience. The expectation that Bunting must be collaborating with the Cuban government added to the audience's curiosity. But the second attempt also stalled, and after a few minutes the screen returned the same error message. Feigning confusion Bunting concluded his presentation and left the stage.

What may not have been immediately clear to the audience was that Bunting's presentation was in fact a performance. He deliberately attempted to load nonexistent Web pages—artwork that didn't exist at all—in a radical expression of solidarity with the network itself. No art object, Web page or otherwise, was necessary for Bunting. Letting the artwork disappear was the very means by which the audience could *experience* the network protocols themselves.

Bunting writes that he was making a point about "the end of net.art," and thus the presentation of nonexistent Web addresses was akin to making a presentation about "no existing work."[29] And Bunting's premonition was correct, for online art making gradually shifted in the late 1990s from being focused on the network (net.art) to being focused on various commercial contexts such as the software industry.

Internet Art as Art of Software

If the birth of net.art can be tied to an email received by Vuk Ćosić in December 1995, then the symbolic starting point for the second phase of In-

---

29. Heath Bunting, personal correspondence, September 20, 2000.

ternet art—a phase I suggest is tempered not by the interests of the network, but by the more commercial interests of the software industry—was January 25, 2000, when the lawsuit against Swiss artists Etoy was withdrawn, signaling the end of the *Toywar,*[30] a two-month global art event that Etoy describes as "the single most expensive performance in art history: $4.5 billion in damage!"[31] *Toywar* was an online gaming platform playable simultaneously by multiple users around the world. The goal of the game was to negatively affect specific capital valuations on the NASDAQ stock market. *Toywar* went on to receive an honorable mention in the Prix Ars Electronica awarded annually at the Ars Electronica festival in Linz.

While corporate organizations have long aestheticized their moneymaking practices in the realm of culture—in everything from Nike advertisements to Giorgio Armani's 2000 exhibition at the Guggenheim Museum—it is only recently that artists have reversed this process and started to aestheticize moneymaking practices in the corporate realm.[32] Taking a cue from the

---

30. On January 25, 2000, an unofficial settlement was signed by eToys and Etoy. The lawsuit was officially dismissed on February 16, 2000. See http://www.toywar.com.

31. Cited from the liner notes to the audio CD *TOYWAR.lullabies* (Etoy 2000).

32. While I argue that a distinct trend has only recently appeared, several historical referents exist as exceptions to the rule. More recent examples include Ingold Airlines, a fictional airline created by Swiss artist Res Ingold, and NSK (Neue Slowenische Kunst), a corps of Slovenian artists who issue passports and engage in other pseudo-commercial state activities. Earlier in the century, Yves Klein, in moments of anti-commercial irreverence, would consummate "transactional" art pieces by flinging gold dust into the Seine River. Other artists have used money as the content of their work including conceptualist Chris Burden, who in the early 1970s received an artist's award from the Los Angeles County Museum of Art. He cashed the check for two hundred $10-bills, which he then individually mailed (inside Christmas cards) to his list of top art world insiders. American artist J. S. G. Boggs gained notoriety for his detailed drawings of money. "Arte-Reembolso" (Art Rebate), a public art project by David Avalos, Elizabeth Sisco, and Louis Hock in San Diego, California, in 1994, also used money as the content of the art performance. In this piece, $10 bills were given out to 450 illegal aliens at the U.S.–Mexico border. The piece was funded by San Diego's Museum of Contemporary Art. Finally, in the most extreme example prior to *Toywar,* Jimmy Cauty and Bill Drummond of the K Foundation (formerly of the pop music group KLF) burned £1 million in Scotland on August 23, 1994. The act is documented in the film *Watch The KLF Burn A Million Quid.* I

corporate sensibilities of art factory production models in the 1980s and the gusto of the late-nineties software industry, artist groups like Etoy and RTMark have begun to think and act like corporations, even going so far as to create mutual funds and issue stocks *as art objects*.

RTMark[33] is a corporation dedicated to anticorporate sabotage activities. It was instrumental in several now famous actions such as the Barbie Liberation Organization in the 1980s, the *Deconstructing Beck* CD from the late 1990s, and also the *Toywar* activities of December 1999. They were featured in the 2000 Whitney Biennial Exhibition in New York.

RTMark is a corporation for practical reasons. Being a corporation displaces liability for culturally subversive and sometimes illegal work. While the artists often dress in corporate attire and give presentations that would be more at home in the boardroom than the museum gallery, it is not simply in uniform that RTMark resembles a corporation. It operates very much like a financial services institution, offering a range of investment products to consumers. Whereas a commercial bank has a range of capital receptacles, from high-tech funds to IRAs, RTMark offers a series of funds that represent different fields of subversive cultural production. For example, the "Media Fund," managed by writer Andrei Codrescu, focuses on acts of corporate sabotage in the mediascape. Invest in culture, says RTMark, not capital.

Like RTMark, the Bureau of Inverse Technology (BIT) is a corporate art production entity. BIT proudly identifies itself as a full-service agency for production, marketing, and commentary, revealing a critical cynicism about the political fabric of techno-products and the persistent lack of "transcendent poetics" in these products. "The cultural force of products frame[s] how we work, how we incorporate nonhuman agency in the mundane daily interactions that form human habit, which then gets called human nature," the agency writes. "The Bureau produces a brand legacy and brand story with ambitions not unlike Nike and Disney. Brand loyalty for the sophisticated consumer is produced through heterogeneous networks of material and

---

thank Benjamin Weil, David Ross, Kerry Tribe, and Jennifer Crowe for bringing some of these artworks to my attention.

33. See http://www.rtmark.com. The name (pronounced ART-mark) is a pun using both the registered symbol (®) and the trademark symbol (TM).

ephemeral culture in which products are embedded. Technoart engages this, unwittingly or not."

Similarly, in her early Internet art project *Bodies INCorporated*[34] Victoria Vesna investigated both corporate business practices and the corporeal body, playing on the meaning of the word "incorporate." In the virtual world of *Bodies INCorporated,* users earn shares based on how involved they are in the various activities and options offered to them. Consequently, more shares allow for greater participation in the community of body owners.

By far the most successful corporate artists are the Swiss art group Etoy.[35] Since 1995 it has won several awards and has received extensive international media attention. As an artist group that is also a corporation, it issues what are called "etoy.SHARES." The shares represent ownership in the company and operate similarly to capital ownership in the stock market system. Etoy.SHARES have a monetary value and can be bought directly from the corporation. Upon receiving an investment from the "client" (the art collector), Etoy issues an original stock certificate printed on aluminum and rendered unique by an embedded "smart chip."[36] The value of etoy.SHARES is recorded on a stock chart maintained by the organization. The rise and fall of the share value corresponds directly to the relative success or failure of the art group in the cultural arena. The etoy.SHARES represent the cultural capital associated with the art group at any given moment. The cultural dividend returned by the artists grows and shrinks in accordance with the share value.

In 1999 Etoy was sued by the Internet toy retailer eToys, who claimed that Internet toy buyers might be confused and potentially offended by the artists's Web site if they typed E-T-O-Y into their Internet browsers rather than E-T-O-Y-S. Since the artists had been using the name well prior to the toy retailer, many in the art world were upset by the lawsuit. The pro-Etoy position received extensive grassroots support from thousands of Internet users including civil liberties advocate John Perry Barlow and author Douglas Rushkoff. The press was also very supportive of the anti-eToys campaign.

---

34. See http://www.arts.ucsb.edu/bodiesinc.

35. See http://www.etoy.com.

36. A collection of Etoy stock certificates were exhibited for the first time in New York in the spring of 2000 at Postmasters Gallery.

War

If the birth of net.art can be tied to an email received by Vuk Ćosić in December 1995, then the symbolic starting point for the second phase of Internet art was January 25, 2000, when the lawsuit against Swiss artists Etoy was withdrawn, signaling the end of the *Toywar*, a two-month global art event that Etoy describes as "the single most expensive performance in art history: $4.5 billion in damage!"

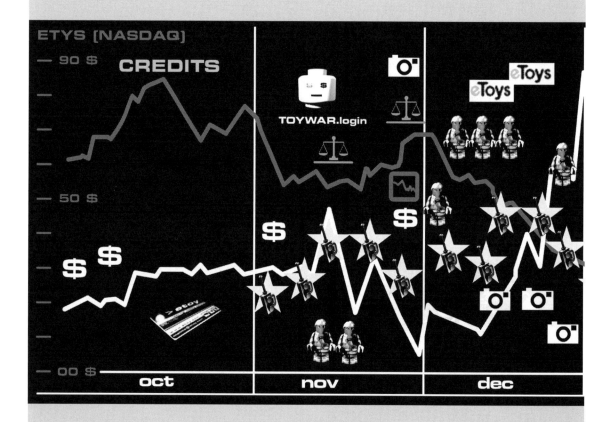

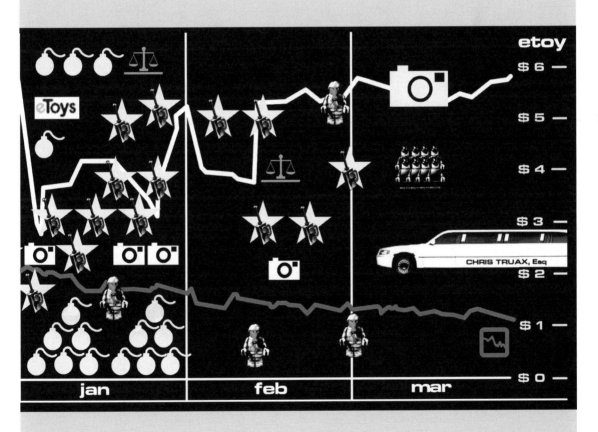

Etoy, *TOYWAR.timeline* (2000)

But this was no ordinary anticorporate protest. Etoy itself did something that was truly breathtaking. It created a piece of software called *Toywar,* which was an online gaming platform for multiple users. "We will release sort of an action entertainment game," the Etoy press speaker announced in December 1999. "People are part of a huge battlefield, where they can fight against eToys Inc. People will have their own character and will have an income—if they work hard they get etoy.SHARE options. They will also decide what's going on—what the next step will be, because the shareholders will decide whether to sell etoy or not."[37] The *Toywar* battlefield is a complex, self-contained system, with its own internal email, its own monetary system, its own social actors, geography, hazards, heroes and martyrs. The goal of *Toywar* was to wage "art war" on eToys Inc., trying to drive its stock price to as low a value as possible—and in the first two weeks of *Toywar,* eToys' stock price on the NASDAQ plummeted by over 50 percent and continued to nosedive. The corporate efficiency and energy of Etoy, who itself would rather disband than part with its dot-com domain name (as this is the very core of its corporate artistic identity), had now been inverted and redirected to another commercial entity, creating what may indeed have been the most expensive performance piece in art history. The strategy worked. eToys Inc. dropped its lawsuit against the artists and declared bankruptcy in 2001.

During *Toywar* art making changed a great deal. Not only did whole new aesthetic sectors open up for art making (in particular, the unprecedented artistic practice of destroying capital on international stock markets), but also the nature of Internet art itself shifted from being defined by the limitations of the network (seen in the earlier net.art movement) to being defined more by the commercial interests of the software industry. This is an incredibly rich moment, a moment that sits nicely in a larger history of avant-garde desires, yet that discovers new aesthetic possibilities specific to the protocols of new media.

Like the struggle in the software industry between proprietary technologies and open, protocological ones, Internet art has struggled between an

37. Etoy, personal correspondence, December 1999. "Selling Etoy" is a reference to the half-million-dollar price tag that eToys offered Etoy to purchase its name. Etoy had declined to sell at that price, prompting the subsequent lawsuit.

aesthetic focused on network protocols, seen in the earlier work, and an aesthetic focused on more commercial software, seen in the later work.

Auctionism

One particular subgenre of Internet art that mixes both sides of the aesthetic divide (art as network and art as software) in interesting ways is auction art. Auction art is any art that uses online auction Web sites like eBay, the Internet's leading auction Web site where scores of articles are auctioned off over the Web every minute of the day. As Robert Atkins writes on the subject: "After Yahoo!, eBay, the online auction site, may be the most well known corporate enterprise on the web. (The four and a half-year-old 'online trading community' currently features 4.4 million items in 4,320 categories on its site.)"[38] eBay art is therefore primarily a commercial art genre in that it engages with the context of buying and selling via the Web. So, eBay art should be considered in the same category as video game art, or software art, or other related commercial genres. But eBay art is also a way of aestheticizing the network itself, and network relations. The actual Web page on eBay is important, but other related places and events are important too, such as the email lists to which the artist posts the announcements of his or her auction, and the interest brought on by the bidding war. The communal network or social space created by the auction art piece supplements the artist's eBay Web page.

The art world first used eBay for very practical tasks such as selling artworks or other artifacts. For example, Wolfgang Staehle of The Thing[39] in New York tried in April 2000 to auction a Web interface taken from The Thing's Web site.[40] In an interview with Baumgärtel, Staehle notes that "the buyer aquires the old Thing interface and domain exclusively. The art projects [hosted on The Thing] are nonexclusive. I feel it is important that the whole project will be preserved in its original context and accessible to the future scholars, historians, etc. What you buy is—in a sense—a bit of

---

38. Robert Atkins, "Art as Auction," available online at http://www.mediachannel.org/arts/perspectives/auction.

39. See http://www.thing.net.

40. See Wolfgang Staehle, "THE THING 4 SALE," *Rhizome,* April 23, 2000.

history."[41] (The reserve price was not met during the auction and the interface was not sold.)

Another example involved the Electronic Disturbance Theater. The EDT also tried to use the auction context in a practical or nonartistic way, to auction off an archive of its work from the period from 1997 to 1999.[42] The artists set a minimum bid price of $8,000. Whoever won the bid would receive a Web archive of its work either on CD-ROM or as a data file. In another example, Staehle used eBay to field donations for the Etoy legal defense fund in December 1999.[43]

These are all examples of the nonartistic uses of eBay by members of the art community. But the auction Web site has also been used as an actual medium for art making or otherwise artistic interventions. In the first recorded usage of the medium, Washington, DC–based artist Jeff Gates sold his personal demographics using the eBay Web site. His announcement read: "Information. The currency of the new millennium! You have it? You want it? What are you willing to pay for it? I'm selling my personal demographics to the highest bidder! What more perfect place to do so but on Ebay, the world's largest flea market."[44] In an interview with Laura McGough, the artist admits: "Everyone wants to know who I am! Information is the commodity of the new millennium. Facts about me, my family, our likes and dislikes, and what we eat, drink, and buy are bought and sold at lightning speeds."[45]

In a similar work, young artist Michael Daines auctioned off his body to the highest bidder during an eBay auction in May 2000.[46] To announce the

---

41. Wolfgang Staehle, "W. Staehle on The thing Sale," *Nettime,* May 12, 1999.

42. The material under auction was originally housed at http://www.thing.net/~rdom. See my "Auction Art Resource List," *Rhizome,* July 19, 2000.

43. The original eBay URL was http://cgi.ebay.com/aw-cgi/eBayISAPI.dll?ViewItem&item =219249164. See Wolfgang Staehle, "contribute to the etoy defense fund," November 13, 1999.

44. Jeff Gates, "Artist Sells Himself on Ebay!," *Rhizome,* June 1, 1999.

45. See Laura McGough, "ebay.art," available online at http://www.nomadnet.org/massage6/ ebayart.

46. See Michael Daines, "The Body of Michael Daines," *Rhizome,* April 25, 2000. The original eBay webpage, now offline, is archived at the artist's Web site. See http://mdaines.com/body.

auction, the artist simply wrote: "now showing at eBay: The body of a 16 year old male. Overall good condition with minor imperfections."[47]

In spring 2000, RTMark, a participating artist in that year's Biennial exhibition at the Whitney Museum in New York, auctioned off its tickets to the Biennial party (hot commodities, available only to the art world "A" list).[48] In this way, the winning bidder would also be able to participate in the Biennial despite not originally being invited, attending the VIP party in lieu of the original artist. In an email announcement for the artistic intervention, RTMark wrote: "Offered: four artist tickets (for two) to the Whitney Biennial's exclusive VIP party for patrons, curators, and Biennial artists only. Winning bidders will become official Whitney Biennial artists."[49] RTMark's artwork in the Biennial was similar in spirit. It allowed its Web site to be hijacked during the course of the exhibition by letting anyone submit a URL that would then replace the original RTMark Web site. Any Web artist could, then, participate in the Biennial for a few minutes or even a day or two by including his or her URL within the framework of the RTMark Web site.

In a similar intervention, the AKSHUN collective at the California Institute of the Arts in Valencia auctioned off a block of time in the school's main gallery space to the highest bidder: "Opening bid—$5.00. More than 300 feet of wall space available for exhibition at prestigious art institute from December 11–December 16, 1999. White walls, track lighting, 24 hour access and security, excellent condition. Located in institute's main building, near central entryway. Guaranteed audience of no less than 1200 art critics, curators and artists."[50] They received bids and inquiries from around the world. The winning bid was a Fluxus exhibition entitled "KLONDIKE: International Fluxus Group Show" organized by Alan Bukoff that hung in the Main Gallery of CalArts from December 12 to December 18, 1999.

---

47. Daines, "The Body of Michael Daines."

48. This piece is partially archived at http://rtmark.com/more/whitneyebay.html.

49. RTMark, "Whitney Biennial Artist Tickets and Status for Sale on eBay," *Rhizome,* March 9, 2000.

50. AKSHUN, "73,440 Minutes of Fame!," *Nettime,* October 15, 2000.

The Body Of Michael Daines

Item #317865927

Antiques & Art:Art:Fine:Sculptures

Description

Bid!

Watch this
item

Currently	**$5.00**
Quantity	**1**
Time left	**9 days, 0 hours +**
Started	Apr-25-00 16:08:53 PDT
Ends	May-05-00 16:08:53 PDT
Seller (Rating)	**emdehns** (1)
	(view comments in seller's Feedback Profile) (view seller's other auctions)
	(ask seller a question)
High bid	--
Payment	Money Order/Cashiers Checks, COD (collect on delivery), Personal Checks
Shipping	Buyer pays actual shipping charges, Will ship to United States and the following regions: Canada
Update item	**Seller:** If this item has received no bids, you may revise it. Seller revised this item before first bid.

First bid	**$5.00**
# of bids	**0** (bid history) (with emails)
Location	**Calgary, AB**
Country	**Canada**
	(mail this auction to a friend)
	(request a gift alert)

Seller assumes all responsibility for listing this item. You should contact the seller to resolve any questions before bidding. Auction currency is U.S. dollars ($) unless otherwise noted.

Description

The body of a 16 year old male.

Overall good condition with minor imperfections.

The Body Of Michael Daines
Item #317865927

Opening bid: $5.00

Your maximum bid: [   ]

(Minimum bid: $5.00)

(Review bid)

eBay will bid incrementally on your behalf **up to your maximum bid**, which is kept secret from other eBay users. The eBay term for this is proxy bidding.

Your bid is a contract - Place a bid only if you're serious about buying the item. If you are the winning bidder, you will enter into a legally binding contract to purchase the item from the seller.

How to Bid

1. Register to bid - if you haven't already. It's free!

2. Learn about this seller - read feedback comments left by others.

3. Know the details - read the item description and payment & shipping terms closely.

4. If you have questions - contact the seller emdehns *before* you bid.

5. Place your bid!

eBay purchases are insured.

Top Questions From This Page

- How do I place a proxy bid? It looks like I can only place a maximum bid.
- Why doesn't my bid show up?
- What does "reserve not yet met" mean?
- How can I change something or cancel my listing completely?
- Why isn't my picture showing up?
- As a seller, how can I cancel an unwanted bid?
- Why does my email address appear when I have a User ID?
- How do I register?

Announcements | Register | SafeHarbor (Rules & Safety) | Feedback Forum | About eBay

TRUST**e**
site privacy statement

Michael Daines, *The Body of Michael Daines* (2000)

Other artists are also working in this genre. New York artist Cary Peppermint has used eBay on several occasions, including the project "Use Me As A Medium."[51] Peppermint is cited as writing: "The high-bidder will have the rare opportunity to make art using artist Cary Peppermint . . . High-bidder will email specific instructions/directions for Cary to perform. High-bidder will then receive a 5–15 minute VHS of Peppermint following high-bidder's instruction."[52] The winning bidder would therefore have complete control over the artist's body, albeit only at a distance, and for a limited period of time.

Auction art, or simply "auctionism," thus exhibits both characteristics of Internet art as I described it earlier. Auctionism unravels the limitations of the network by moving the location of the art object off the Web site and into the social space of the Net, particularly email lists like *Rhizome, Nettime,* and others. It is a performance that exists both on eBay and also on the email lists where the piece is advertised. Communities of email list subscribers comment and bid on the artwork, making it a type of social exchange. But at the same time, auctionist art operates within the limitations of the commercial genre of the online auction, thereby showing what I describe as the second phase or theme of Internet art. In conclusion, let me restate the periodization that I suggest helps one understand Internet art practice from 1995 to the present day. The early, conceptual phase of Internet art known as "net.art" is concerned primarily with the network, while the later, corporate phase is concerned primarily with software.

---

51. This piece is archived at http://www.restlessculture.net/peppermint/exposures/images/me.gif.

52. Cited in Robert Atkins, "Art as Auction," available online at http://www.mediachannel.org/arts/perspectives/auction.

Conclusion

Let me end with a hypothetical example. On an empty town street where drivers are prone to speed, the local residents elect to have speed bumps installed. The inhabitants assume that the speed bumps will reduce the amount of speeding through their community.

In another part of town is an identical street also plagued by speeding. But here the neighborhood decides against speed bumps. Residents pass laws to reduce the legal speed limit. They install speed limit signs and increase police radar surveillance. Both solutions succeed in reducing the number of speeding cars. But which solution is *protocological?*

Many would guess the second solution is the protocological one. It appears to be the most current, the most technological, and in some ways the most politicized and insidious.

But it is the former. Whereas signage and police compel the driver to slow down, they do so through what amounts to nothing more than a polite request. Yes, there is always the threat of violence, but essentially the burden of slowness is placed onto the decision-making process of the speeding driver who may or may not agree to abide by the rules (and accept the consequences).

Bumps, on the other hand, create a physical system of organization. They materially force the driver to acquiesce. Driving slower becomes advantageous. With bumps, the driver *wants* to drive more slowly. With bumps, it becomes a virtue to drive slowly. But with police presence, driving slowly can never be more than coerced behavior. Thus, the signage appeals to the mind, while protocol always appeals to the body. Protocol is not a superego (like the police); instead it always operates at the level of desire, at the level of "what we want."

This book is divided into three parts. Part I examines protocol in its positive representation, specifically how protocol exists as a physical technology, as a formal technology, and also as a political technology. Parts II and III examine protocol in its negative representation, a profile that Armand Mattelart describes well within a larger context of globalization:

When some political scientists speak of the "new global fronts of disorder," "areas of darkness," or "anti-worlds," they are referring to fundamentalism, sects, channels or the underground of informal economy, Mafia networks and illicit trafficking (from narcotics to children or contraband electronics), transnational flows of diasporas and migrant labor—both regular and illegal—toward the affluent countries and regions, etc. These dissonant fronts and parallel worlds reveal the crises, conflicts, and imbalances affecting our changing societies and confront them [with] the constant risk of collapse or disaster.[1]

This constant risk of collapse or disaster is what makes the subcultures discussed here—hacking, cyberfeminism, Net art—so necessary for the assumption and continuing maturation of protocol.

Protocol is a management style that injects control into "fronts of disorder," "anti-worlds" (whatever those are), "Mafia networks," and "crises, conflicts, and imbalances." Today, protocol is our gravity, our oxygen, our pulse.

Historically the relaxing of hierarchy in social and political systems presented a management problem: How can complete chaos be avoided when all traditional systems of control (hierarchy, centralization, etc.) have slipped away? Protocol is a solution to the problem of hierarchy. It is in many ways an historical advancement.

"Many systems are organised hierarchically" like trees are, writes Berners-Lee. "A tree has the practical advantage of giving every node a unique name. However it does not allow the system to model the real world."[2] Hierarchy may allow for greater efficiency or greater instrumentalism, but as Berners-Lee points out, it is less faithful to the actual material existence of the real world today. As Deleuze and Guattari put it simply, "We're tired of trees."[3]

Yet the success of protocol today as a management style proves that the ruling elite is tired of trees too. "One essential characteristic that sets late capitalism apart from other political and economic forms," writes Critical Art Ensemble on the disappearance of hierarchical power, "is its mode of representing power: What was once a sedentary concrete mass has now become a nomadic electronic flow."[4] So the rich and powerful are also profiting from the transition into protocological control.

---

1. Armand Mattelart, *Networking the World, 1794–2000* (Minneapolis: University of Minnesota Press, 2000), p. 101.

2. Tim Berners-Lee, *Weaving the Web* (New York: HarperCollins, 1999), p. 215.

3. Gilles Deleuze and Félix Guattari, *A Thousand Plateaus* (Minneapolis: University of Minnesota Press, 1987), p. 15.

4. Critical Art Ensemble, *Electronic Civil Disobedience and Other Unpopular Ideas* (New York: Autonomedia, 1996), p. 7.

But power's enemies are swimming in that same flow. Media historians Randall Packer and Ken Jordan gleefully proclaim that "[m]ultimedia, by its very nature, is open, democratic, nonhierarchical, fluid, varied, inclusive."[5] Their claim is essentially true, but for completely opposite reasons than they might hope for. I aim here to answer the question: How is a technology able to establish real-world control when it lacks certain fundamental tools such as hierarchy, centralization, and violence? Why does technology seem, as Kevin Kelly likes to put it, so "out of control" yet still function so flawlessly? There must be some machine that, at the end of the day, sorts it all out.

Protocol is that machine, that massive control apparatus that guides distributed networks, creates cultural objects, and engenders life forms.

This book is not meant to be a history of protocol or protocols. Protocol has its own extensive history in diplomacy, the military, the government, and the private sector. Instead, I offer a new story of protocol as it intersects with both the digital computer and the distributed network, two historically specific technologies. This three-way historical intersection coincides with the start of the new millennium and therefore will affect culture for many years to come.

Let me end by restating a few summarizing moments selected from previous chapters.

- Protocol is a system of distributed management that facilitates peer-to-peer relationships between autonomous entities.
- Internet protocols allow for inter-operation between computers.
- Protocol's virtues include robustness, contingency, inter-operability, flexibility, heterogeneity, and pantheism.
- A goal of protocol is totality. It must accept everything, no matter what source, sender, or destination. It consumes diversity, aiming instead for university.
- Protocol is a universalism achieved through negotiation, meaning that in the future protocol can and will be different.
- Facilitated by protocol, the Internet is the mostly highly controlled mass media hitherto known.

---

5. Randall Packer and Ken Jordan, eds., *Multimedia: From Wagner to Virtual Reality* (New York: Norton, 2001), p. xxx.

- Protocol is materially immanent, but protocological objects never contain their own protocol. Protocols generally resist interpretation.
- Protocol is a language that regulates flow, directs netspace, codes relationships, and connects life forms. It is etiquette for autonomous agents.
- Self-determinism of material systems is a precondition of protocol.
- Protocol is a type of controlling logic that operates largely outside institutional, governmental, and corporate power.
- In order to be politically progressive, protocol must first be partially reactionary.
- The best tactical response to protocol is not resistance but hypertrophy.
- The current global crisis is one between centralized, hierarchical powers and distributed, horizontal networks. But in the future one is likely to see bilateral organizational conflict, that is, networks fighting networks.
- Code is the only language that is executable, meaning that it is the first discourse that is materially affective.
- Protocol is synonymous with possibility.
- Tactical media are effective at exploiting flaws in protocological technologies.

As the previous few chapters show, the assumption of protocol as today's dominant management style is far from complete and, in reality, quite vulnerable. It is very likely if not inevitable that the core Internet protocols, today largely safe from commercial and state power, will be replaced by some type of proprietary system. (The fact that Microsoft has not yet replaced TCP/IP with a commercial product of its own is one of the miracles of computer history. Chances are this will happen very soon.)

As one learns more and more about the networks of protocological control, it becomes almost second nature to project protocol into every physical system: Traffic lights become the protocol for successful management of moving vehicles; a grocery store queue is the protocol for a successful checkout; airport security points are the protocol for prohibiting weapons; and so on. Protocol pops up everywhere.

But protocol is more than simply a synonym for "the rules." Instead, protocol is like the trace of footprints left in snow, or a mountain trail whose route becomes fixed only after years of constant wear. One is always free to pick a different route. But protocol makes one instantly aware of the best route—and why wouldn't one want to follow it?

Thus, a better synonym for protocol might be "the practical," or even "the sensible." It is a physical logic that delivers two things in parallel: the solution to a problem, plus the background rationale for why that solution has been selected as the best. Like liberalism, or democracy, or capitalism, protocol is a successful technology precisely because its participants are evangelists, not servants. Like liberalism, democracy, or capitalism, protocol creates a community of actors who perpetuate the system of organization. And they perpetuate it even when they are in direct conflict with it.

Protocol then becomes more and more coextensive with humanity's productive forces, and ultimately becomes the blueprint for humanity's innermost desires about the world and how it ought to be lived.

This makes protocol *dangerous*—but in the Foucauldian sense of danger that is twofold. First it is dangerous because it acts to make concrete our fundamentally contingent and immaterial desires (a process called reification), and in this sense protocol takes on authoritarian undertones. As a colleague Patrick Feng said recently: "Creating core protocols is something akin to constitutional law," meaning that protocols create the core set of rules from which all other decisions descend. And like Supreme Court justices having control over the interpretation of the American Constitution, whoever has power over the creation of such protocols wields power over a very broad area indeed. In this sense protocol is dangerous.

But protocol is also dangerous in the way that a weapon is dangerous. It is potentially an effective tool that can be used to roll over one's political opponents. And protocol has already proven this in the sphere of technology. What poses a real threat to Microsoft's monopoly? Not Macintosh (the market). Not the Justice Department (the state). Instead it is the widespread use of protocols that struggle against Redmond's proprietary standards with varying degrees of success. What poses a real threat to global dominance of American culture? Not the French Ministry of Culture. Not Bollywood. Instead it is the transitory networks and temporary autonomous zones—each protocological spaces—that bring out vulnerabilities where before none were thought to exist.

People often ask me if I think protocol is good or bad. But I'm not sure this is the best question to ask. It is important to remember first that the technical is always political, that *network architecture is politics*. So protocol necessarily involves a complex interrelation of political questions, some progressive, some reactionary. In many ways protocol is a dramatic move forward, but in other ways it reinstates systems of social and technical control that are deserving of critical analysis. Perhaps an analogy from Berners-Lee would help clarify this. The Web, he writes, is like a market economy:

In a market economy, anybody can trade with anybody, and they don't have to go to a market square to do it. What they do need, however, are a few practices everyone has to agree to, such as the currency used for trade, and the rules of fair trading. The equivalent of rules for fair trading, on the Web, are the rules about what a URI means as an address, and the language the computers use—HTTP—whose rules define things like which one speaks first, and how they speak in turn.[6]

With this analogy in place, then, a critique of protocol becomes clearer. In many ways market economies represent a dramatic leap forward in the history of mankind, for they represent a higher degree of individual freedom over previous social forms (e.g., feudalism). But at the same time market economies bring into existence high levels of social inequality. Berners-Lee's "currency used for trade" is clearly not accessible in equal measures for all parties involved, and his "rules for fair trading" have historically been less than fair, subjugating vast populations of the working poor, immigrants, women, and workers in developing countries, to name a few examples.

Thus the same types of critiques that can be levied against so-called successful social realities such as market economies (or even liberalism, or civil society, or the bourgeois class itself) can be levied against protocol. As critics we must first ask ourselves: Do we *want* the Web to function like a market economy? Can we imagine future technological solutions that fulfill our social desires more fully than protocol can? One hopes that protocol's mandate toward openness will help rather than hinder us in implementing these social desires in the future. My goal here in this book has been not to come down cleanly and say that protocol is either good or bad—because clearly it is both, in varying degrees and contexts—but rather to chart some of protocol's distinctive characteristics so that any concerned person can better decide how to evaluate and critique a given protocological technology in a specific historical context.

---

6. Berners-Lee, *Weaving the Web,* p. 36.

Alexander R. Galloway is Assistant Professor of Media Ecology at New York University. Galloway previously worked for six years as Director of Content and Technology at Rhizome.org. He is a founding member of the software development group RSG, whose data surveillance system "Carnivore" was awarded a Golden Nica in the 2002 Prix Ars Electronica. He lives in New York City.

Index

Hypertext Markup Language (HTML),
26, 51–52, 65, 76–77, 137–138,
141, 218, 225
Hypertext Transfer Protocol (HTTP),
51–52, 137–138, 216

Identity, 66, 69, 113–114, 188, 194
I Love You (virus), 187
Inclusion, 142, 147
Incorporations, 103
Information, 72–73, 104, 139
Information age, 111
"Information" (exhibition), 210n7
Information science, 110–111
Information Age, The (Castells), 24
Ingold Airlines, 227n32
Innen, 195
Institute of Contemporary Arts (London), 210n7
Institute of Electrical and Electronics
Engineers (IEEE), 122, 126–127, 130
Institute of Radio Engineers (IRE), 126
Intel, 121
Interface, 191
"Interior Scroll" (Schneemann), 195
Intermedia network (Youngblood), 78
International Electrotechnical Commission (IEC), 129–130
International Organization for Standardization (ISO), 39, 124, 129–130
International Telecommunications
Union (ITU), 129–130
Internet, 29–30
as distributed network, 4, 38
layer (*see* Layer, Internet)
origins of, 4, 29
as sculpture, 216–217
as standard, 6
Internet Architecture Board (IAB),
131–134

Internet Activities Board. *See* Internet
Architecture Board
Internet art, 209–238
corporate aesthetics of, 219, 227–232
net.art, 211–226, 233–238
Internet Corporation for Assigned
Names and Numbers (ICANN),
121, 131n24, 141
Internet-Draft, 134–136
Internet Engineering Steering Group
(IESG), 131–135
Internet Engineering Task Force
(IETF), 6, 130–134, 138, 142
definition of, 133
Internet Protocol (IP), 5, 8, 26, 44–46,
124, 136–137, 143, 244
address, 65
Internet Research Task Force (IRTF), 132
Internet Society (ISOC), 6, 131–132
Internet Standard, 134–136, 140
Interpretation, 52, 139
Interstate Highway System, 35, 38
Ippolito, Jon, 201, 204
Irigaray, Luce, 188

Jameson, Fredric, 23–24, 168
Java, 108. *See also* Programming
Javascript, 220
Jencks, Charles, 26
Jeremijenko, Natalie, 16n24
Jerusalem virus, 179–180
Jevbratt, Lisa, 217
Jewish Museum (New York), 210n7
Jodi, 211, 216–217, 219–224
Johnson, Steven, 221
Joint Technical Committee, 130
Jordan, Ken, 243
Joy, Bill, 124
Judicial system, U.S., 31
JVC, 125–126

Openness, 128, 136, 142

Open source, 171

Open Systems Interconnection Reference Model, 39, 129–130

Operating system, 221

Organic, 93, 96

OSS (Jodi), 221, 224

Outlook. *See* Microsoft Outlook

Packer, Randall, 243

Packet-switching, 5

Palestine Liberation Organization (PLO), 180

Panama, 181

Pandemonium (Hookway), 18, 110

Panopticon, 13, 31

Patrick, Neal, 153

PC Business World, 180

Peer-to-peer communication, 46

Peppermint, Cary, 238

Performance, 226–227, 232

Perl, 167

Peters, Tom, 158–159

Petzold, Charles, 81, 105n81

Phrack, 153, 159, 161

Pierce, Julianne, 192–193

Plane of Immanence (Deleuze), 105

Plant, Sadie, 19, 185, 188–191, 193, 195–196

Popp, Joseph, 181

Possibility, 52–53, 56, 167–172, 210, 244

Postel, Jon, 121–122, 137

Postmodern, 20, 24, 108, 158

"Postscript on Control Societies" (Deleuze), 3, 12, 21, 86. *See also* Society, control

Powell Doctrine, 205n72

Privacy, 113

Programming
assembly, 166–167
language, 65 (*see also* Language)
object-oriented, 33, 108–109

Proprietary, 121, 126, 172, 175

Protocol, 74–75, 140–143, 241–246
as algorithm, 30
as dangerous, 245
and distributed networks, 38
etymology of, 7, 29
failures of, 120
as formal apparatus, 55
immanence of, 50
and life forms, 17
logic of, 212
as management style, 3, 242
nested, 10
as object, 74
and patriarchy, 189–190
politics of, 81
as sculpture, 216–217
as standard, 39

Quittner, Joshua, 161

Radai, Yisrael, 179–180

Radio, 55–56
Net, 211

Rajchman, John, 88

Rakavy, Yuval, 179

Rand Corporation, 4–5, 35, 127, 140n43, 200n65, 204

Ray, Tom, 109, 160, 178

_readme (Bunting), 225

Reaper, 182

Recode, 215

Record, 72

Redcode, 182

Refresh (Shulgin), 215–216